RUNAWAY SCIENCE

TRUE STORIES OF RAGING ROBOTS AND HI-TECH HORRORS

NICK REDFERN

About the Author

Nick Redfern has written more than 60 books. He works full time as an author, lecturer, and journalist. He writes about a wide range of unsolved mysteries, including Bigfoot, UFOs, the Loch Ness Monster, alien encounters, and government conspiracies. His many books include *Nessie*, *The Real Men in Black*, *Shapeshifters*, *Monsters of the Deep*, *The Alien Book*, *The Zombie Book*, *The Bigfoot Book*, *The Monster Book*, *Cover-Ups & Secrets*, *Area 51*, *Secret History*, and *The Roswell UFO Conspiracy*. He writes regularly for *Mysterious Universe*. He has appeared on numerous television shows, including History Channel's *Monster Quest*, *Ancient Aliens* and *UneXplained*; VH1's *Legend Hunters*; National Geographic Channel's *The Truth about UFOs* and *Paranatural*; the BBC's *Out of this World*; MSNBC's *Countdown*; and SyFy Channel's *Proof Positive*. Nick lives just a few miles from Dallas, Texas' infamous Grassy Knoll and can be contacted at his blog: *http://nickredfernfortean.blogspot.com.*

RUNAWAY SCIENCE

TRUE STORIES OF RAGING ROBOTS AND HI-TECH HORRORS

NICK REDFERN

Other Visible Ink Press Books by Nick Redfern

Area 51: The Revealing Truth of UFOs, Secret Aircraft, Cover-ups & Conspiracies
ISBN: 978-1-57859-672-0

The Bigfoot Book: The Encyclopedia of Sasquatch, Yeti, and Cryptid Primates
ISBN: 978-1-57859-561-7

Control: MKUltra, Chemtrails and the Conspiracy to Suppress the Masses
ISBN: 978-1-57859-638-6

The Monster Book: Creatures, Beasts, and Fiends of Nature
ISBN: 978-1-57859-575-4

The New World Order Book
ISBN: 978-1-57859-615-7

Secret History: Conspiracies from Ancient Aliens to the New World Order
ISBN: 978-1-57859-479-5

Secret Societies: The Complete Guide to Histories, Rites, and Rituals
ISBN: 978-1-57859-483-2

The Zombie Book: The Encyclopedia of the Living Dead
With Brad Steiger
ISBN: 978-1-57859-504-4

Also from Visible Ink Press

Alien Mysteries, Conspiracies, and Cover-Ups
By Kevin D. Randle
ISBN: 978-1-57859-418-4

Ancient Gods: Lost Histories, Hidden Truths, and the Conspiracy of Silence
By Jim Willis
ISBN: 978-1-57859-614-0

The Astrology Book: The Encyclopedia of Heavenly Influences, 2nd edition
By James R. Lewis
ISBN: 978-1-57859-144-2

Conspiracies and Secret Societies: The Complete Dossier, 2nd edition
By Brad Steiger and Sherry Hansen Steiger
ISBN: 978-1-57859-368-2

Demons, the Devil, and Fallen Angels
By Marie D. Jones and Larry Flaxman
ISBN: 978-1-57859-613-3

The Dream Interpretation Dictionary: Symbols, Signs, and Meanings
By J. M. DeBord
ISBN: 978-1-57859-637-9

The Government UFO Files: The Conspiracy of Cover-Up
By Kevin D. Randle
ISBN: 978-1-57859-477-1

Haunted: Malevolent Ghosts, Night Terrors, and Threatening Phantoms
By Brad Steiger
ISBN: 978-1-57859-620-1

Hidden Realms, Lost Civilizations, and Beings from Other Worlds
By Jerome Clark
ISBN: 978-1-57859-175-6

The Horror Show Guide: The Ultimate Frightfest of Movies
By Mike May
ISBN: 978-1-57859-420-7

The Illuminati: The Secret Society That Hijacked the World
By Jim Marrs
ISBN: 978-1-57859-619-5

Real Aliens, Space Beings, and Creatures from Other Worlds,
By Brad Steiger and Sherry Hansen Steiger
ISBN: 978-1-57859-333-0

Real Encounters, Different Dimensions, and Otherworldly Beings
By Brad Steiger with Sherry Hansen Steiger
ISBN: 978-1-57859-455-9

Real Ghosts, Restless Spirits, and Haunted Places, 2nd edition
By Brad Steiger
ISBN: 978-1-57859-401-6

Real Miracles, Divine Intervention, and Feats of Incredible Survival
By Brad Steiger and Sherry Hansen Steiger
ISBN: 978-1-57859-214-2

Real Monsters, Gruesome Critters, and Beasts from the Darkside
By Brad Steiger and Sherry Hansen Steiger
ISBN: 978-1-57859-220-3

Real Vampires, Night Stalkers, and Creatures from the Darkside
By Brad Steiger
ISBN: 978-1-57859-255-5

Real Visitors, Voices from Beyond, and Parallel Dimensions
By Brad Steiger and Sherry Hansen Steiger
ISBN: 978-1-57859-541-9

Real Zombies, the Living Dead, and Creatures of the Apocalypse
By Brad Steiger
ISBN: 978-1-57859-296-8

The Religion Book: Places, Prophets, Saints, and Seers
By Jim Willis
ISBN: 978-1-57859-151-0

The Sci-Fi Movie Guide: The Universe of Film from Alien to Zardoz
By Chris Barsanti
ISBN: 978-1-57859-503-7

Supernatural Gods: Spiritual Mysteries, Psychic Experiences, and Scientific Truths
by Jim Willis
ISBN: 978-1-57859-660-7

The UFO Dossier: 100 Years of Government Secrets, Conspiracies, and Cover-Ups
By Kevin D. Randle
ISBN: 978-1-57859-564-8

Unexplained! Strange Sightings, Incredible Occurrences, and Puzzling Physical Phenomena, 3rd edition
by Jerome Clark
ISBN: 978-1-57859-344-6

The Vampire Book: The Encyclopedia of the Undead, 3rd edition
by J. Gordon Melton
ISBN: 978-1-57859-281-4

The Werewolf Book: The Encyclopedia of Shape-Shifting Beings, 2nd edition
by Brad Steiger
ISBN: 978-1-57859-367-5

The Witch Book: The Encyclopedia of Witchcraft, Wicca, and Neo-paganism
by Raymond Buckland
ISBN: 978-1-57859-114-5

"Real Nightmares" E-Books

Book 1: True and Truly Scary Unexplained Phenomenon
Book 2: The Unexplained Phenomena and Tales of the Unknown
Book 3: Things That Go Bump in the Night
Book 4: Things That Prowl and Growl in the Night
Book 5: Fiends That Want Your Blood
Book 6: Unexpected Visitors and Unwanted Guests
Book 7: Dark and Deadly Demons
Book 8: Phantoms, Apparitions, and Ghosts
Book 9: Alien Strangers and Foreign Worlds
Book 10: Ghastly and Grisly Spooks
Book 11: Secret Schemes and Conspiring Cabals
Book 12: Freaks, Fiends, and Evil Spirits

Please visit us at visibleinkpress.com

Runaway Science: True Stories of Raging Robots and Hi-Tech Horrors

Visible Ink Press®
43311 Joy Rd., #414
Canton, MI 48187-2075

Visible Ink Press is a registered trademark of Visible Ink Press LLC.

Most Visible Ink Press books are available at special quantity discounts when purchased in bulk by corporations, organizations, or groups. Customized printings, special imprints, messages, and excerpts can be produced to meet your needs. For more information, contact Special Markets Director, Visible Ink Press, www.visibleink.com, or 734-667-3211.

Managing Editor: Kevin S. Hile
Art Director: John Gouin
Typesetting: Lumina Datamatics Limited
Proofreaders: Larry Baker and Christa Gainor
Indexer: Shoshana Hurwitz

Cover images: Shutterstock.

Cataloging-in-Publication Data is on file at the Library of Congress.

ISBN paperback: 978-1-57859-801-4
ISBN ebook: 978-1-57859-842-7
ISBN hardbound: 978-1-57859-841-0

10 9 8 7 6 5 4 3 2 1

Printed in the United States of America.

Contents

Photo Sources

Asklepioscaduceus (Wikicommons): p. 210.
Marshall Astor: p. 7.
Albert K. Bender: p. 66.
Cabinet des Mádaillesm, Bibliothèque Nationale de France, Paris: p. 15.
Central Intelligence Agency: p. 228.
Computer Designed Organisms (https://cdorgs.github.io/): p. 77.
Creationontheweb.com: p. 260.
DARPA: p. 192.
Davidnoy (Wikicommons): p. 159.
Einsteinsworld.com: p. 178.
Frankie Fouganthin: p. 248.
Future of Humanity Institute: p. 113.
Galaxy Publishing/Ed Emshwiller: p. 27.
JELVi (Wikicommons): p. 154.
Raymond Kurzweil: p. 115.
Simon Leatherdale: p. 63.
Lmdecker (Wikicommons): p. 173.
Los Angeles Times: p. 21.
Tyler Merbler: p. 207.
NASA: pp. 18, 134, 136, 137, 139, 170.
National Institute of Health: p. 160.
Pit-yacker (Wikicommons): p. 32.
Planetkazik (Wikicommons): p. 70.
Daniel Salter (courtesy Nick Redfern): p. 158.
Philippe Semeria: p. 17.
Shutterstock: pp. 4, 6, 24, 28, 36, 43, 45, 48, 50, 52, 57, 73, 75, 79, 81, 83, 86, 90, 97, 104, 105, 108, 127, 128, 129, 142, 145, 164, 167, 168, 180, 184, 200, 204, 216, 217, 220, 224, 230, 236, 240, 243, 249, 250, 254.
Brad Steiger: p. 54.
Thechirurgeonsapprentice.com: p. 265.
Twentieth Century-Fox Film Corp.: pp. 117. 118.
Uffizi Gallery, Florence, Italy: pp. 11, 13.
United Artists: p. 238.
University of Texas Medical Branch at Galveston: p. 162.
U.S. Air Force: pp. 59, 87.
U.S. Congress: pp. 165, 212.
U.S. Department of Defense: p. 193.
U.S. Navy: p. 94.
Jiuguang Wang: p. 1.
White House Photographic Collection: p. 61.
Paul Wicks: p. 198.
Public domain: pp. 8, 112.

Acknowledgments

I would like to offer my very sincere thanks to my tireless agent and agent, Lisa Hagan, and to everyone at Visible Ink Press, and particularly Roger Janecke and Kevin Hile.

Introduction

It was science-fiction legend Isaac Asimov who came up with a series of rules for robots. Of course, Asimov's words were born out of his famous novel *I, Robot*, rather than out of fact. Nevertheless, we can still apply Asimov's words to the real world. They go as follows:

- A robot may not injure a human being or, through inaction, allow a human being to come to harm.
- A robot must obey the orders given it by human beings except where such orders would conflict with the First Law.
- A robot must protect its own existence as long as such protection does not conflict with the First or Second Laws.

As for *I, Robot* itself, it can be understood via the blurb on the cover of the book: "*I, Robot*, the first and most widely read book in Asimov's Robot series, forever changed the world's perception of artificial intelligence. Here are stories of robots gone mad, of mind-reading robots, and robots with a sense of humor. Of robot politicians, and robots who secretly run the world—all told with the dramatic blend of science fact and science fiction that has become Asimov's trademark. With these three, simple directives, Isaac Asimov formulated the laws governing robots' behavior. In *I, Robot*, Asimov chronicles the development of the robot from its primitive origins in the present to its ultimate perfection in the not-so-distant future—a future in which humanity itself may be rendered obsolete."

The story of robots and what we now call "hi tech" go back further in history than you might know. According to a writer for the Stanford University School of Engineering: "One of the first instances of a mechanical device built to regularly carry out a particular physical task occurred around 3000 B.C.E.: Egyptian water clocks used human figurines to strike the hour bells. In 400 B.C.E., Archytus of Taremtum, inventor of the pulley and the screw, also invented a wooden pigeon that could fly. Hydraulically operated statues that could speak, gesture, and prophecy were commonly constructed in Hellenic Egypt during the second century B.C.E.

"In the first century C.E., Petronius Arbiter made a doll that could move like a human being. Giovanni Torriani created a wooden robot that could fetch the Emperor's daily bread from the store in 1557. Robotic inventions reached a relative peak (before the 20th century) in the 1700s; countless ingenius, yet impractical, automata (i.e. robots) were created during this time period. The 19th century was also filled with new robotic creations, such as a talking doll by Edison and a steam-powered robot by Canadians. Although these inventions throughout history may have planted the first seeds of inspiration for the modern robot, the scientific progress made in the 20th century in the field of robotics surpass previous advancements a thousandfold."

CTE Publications add to this saga: "When many Americans think of the word 'robot,' years of science fiction portrayals and action movies immediately come to mind. And while science fiction often misses the mark, the history of robots actually owes quite a debt to science fiction masters like Isaac Asimov. However, to truly understand the history and evolution of robotics, we have to define the term. That's surprisingly difficult to do. For our purposes, we're going to define a robot as a machine that's capable of carrying out routine or complex actions that are programmed by engineers. Today, robots can be used for surgery, massage therapy, space exploration, manufacturing, and code analysis, but the earliest robots were far more primitive—they were tools that could tell time or automotoms that could perform for entertainment. Broadly defined, humans have been developing robotics and automata for hundreds of years. . . ."

The Mind Project has had significant input in this issue, too: "In the 20th century, the digital computer is invented. Researchers quickly start referring to the computer as an 'electronic brain' and start thinking about ways to build robots with computer brains. The first modern programmable robot was the *Unimate.* General Motors installed the first robot to work in a factory in 1961 to move pieces of hot metal. Unimate was an autonomous, pre-programmed robot that repeatedly performed the same dangerous task.

"In 1966, Shakey the Robot is invented at Stanford. Shakey was the first autonmous, intelligent robot that made its own decisions about how to behave. Shakey could be given general instructions, such as "move the block onto the table" and it would reason how to perform the task. This would involve looking around the room, identifying the block and the table, and then figuring out how to get the block to the table, including navigating around any obstacles in the room. In 2004 Shakey was inducted into Carenegie Mellon's Robot Hall of Fame."

ThoughtCo. gets right to the early years of how robots began in the realms of fiction and entertainment: "Writers and visionaries envisioned a world including robots in daily life. In 1818, Mary Shelley wrote 'Frankenstein,' which was about a frightening artificial lifeform come to life by a mad, but brilliant scientist, Dr. Frankenstein. Then, 100 years later Czech writer Karel Capek coined the term robot, in his 1921 play called 'R.U.R.' or 'Rossum's Universal Robots.' The plot was simple and terrifying; the man makes a robot then robot kills a man. In 1927, Fritz Lang's 'Metropolis' was released. The Maschinenmensch ('machine-human'), a humanoid robot, was the first robot ever to be depicted on film.

"Science fiction writer and futurist Isaac Asimov first used the word 'robotics' in 1941 to describe the technology of robots and predicted the rise of a powerful robot industry. Asimov wrote 'Runaround,' a story about robots which contained the 'Three Laws of Robotics,' which centered around Artificial Intelligence ethics questions. Norbert Wiener published 'Cybernetics,' in 1948, which formed the basis of practical robotics, the principles of cybernetics based on artificial intelligence research."

Despite the above, it's a fact that we can go back much further in time, when it comes to the development of the robot—as we shall see right now.

Living Tech

Her name is Sophia. She's the world's most famous robot. She looks just like us. And she's a little bit creepy. Some might say she's *very* creepy. In fact, *many* have said exactly that. Is Sophia just an expensive gimmick? Or possibly a precursor to what our future might look like—that is to say, a world dominated by sophisticated, bullying robots who view us as their underlings? Could such a thing happen? Possibly, the road to a real-life *I, Robot* may not be too far away. Sophia's creators, a Hong Kong–based company, Hanson Robotics, have ensured that Sophia can imitate more than 60 human expressions. It's not just the facial appearances that are changing; the times and the technology are, too.

It should be noted that Sophia, in the real world, parallels a fictional robot from the 1920s, specifically from the movie *Metropolis*.

This is a replica of Maria, the robot featured in the 1927 Fritz Lang movie Metropolis. *It is currently on display at the Carnegie Science Center's Robot Hall of Fame in Pittsburgh, Pennsylvania.*

Roger Ebert's review of the 2010 restoration of *Metropolis* is glowing, and rightly so: "The opening shots of the restored *Metropolis* are so crisp and clear they come as a jolt. This mistreated masterpiece has been seen until now mostly in battered prints missing footage that was, we now learn, essential. Because of a 16mm print discovered in 2008 in Buenos Aires, it stands before us as more or less the film that Fritz Lang originally made in 1927. It is, says expert David Bordwell, 'one of the great sacred monsters of the cinema.'

"Lang tells of a towering city of the future. Above ground, it has spires and towers, elevated highways, an Olympian stadium and Pleasure Gardens. Below the surface is a workers' city where the clocks show 10 hours to squeeze out more work time, the workers live in tenement housing and work consists of unrelenting service to a machine. This vision of plutocracy vs. labor would have been powerful in an era when the assembly line had been introduced on a large scale and Marx had encouraged class warfare."

To say anything else would spoil things for those who have yet to see the production. I will just say that while we, today, have Sophia, *Metropolis* has Maria. Now, let's get back to Sophia.

The London Speaker Bureau provides us with a summary of Sophia and her "life." The bureau says: "Sophia has discussed subjects ranging from 'Will robots take over the world?' to how artificial intelligence [AI] could end hunger in developing nations. Sophia interacts with humans in a profoundly personal way, previewing a fast-approaching future where friendly, caring humanoids help us solve our most challenging problems to create a better world. Sophia the Robot has sung in concert, addressed audiences in Mandarin, debated the future of the human race (against another robot), and generated billions of views and social media interactions. She has also displayed her expansive tech knowledge and growing business savvy, meeting with leaders in the banking, insurance, automotive, property development, media and entertainment industries."

It's no wonder there are people who are concerned that the day may come when we are seen as second-class citizens. And it isn't hard to guess who might eventually become the first-class ones.

Writer Brent Swancer has taken an interest in the phenomenon known as Sophia. He says that "the robot was first turned on in 2016 and is immediately notable for how realistic she (we'll call it 'she') looks. Modeled after the ancient Egyptian Queen Nefertiti, Sophia possesses an incredibly intricate array of servo motors beneath her lifelike "skin," allowing her to emulate more than 60 facial expressions and move her face and eyes in subtle ways to express complex emotional cues. Indeed, at first glance she could almost pass for a real person if it weren't for her lack of a wig on her head, leaving her robotics exposed. With her very realistic countenance and the lack of a covering for the exposed machinery, Sophia is at once amazing, lifelike, and rather unsettling, but this is only the beginning of how impressive and spooky she really is."

"Impressive" is the word. Paul Seaburn, who has followed the ever-developing saga of Sophia, makes that very clear: "Art painted by Sophia, the world's most human-like robot, sold at an auction for $688,888, so now she's working on creating music, complete with lyrics."

Things get even more surreal, as Seaburn demonstrates: "Sophia, the first robot to ever have been granted citizenship in any nation (in her case, by Saudi Arabia), has announced that she wants to have a baby and start a family of little AI princes and princesses." I'd like to see how that would come to fruition. Not only does Sophia have citizenship, she also has her very own passport.

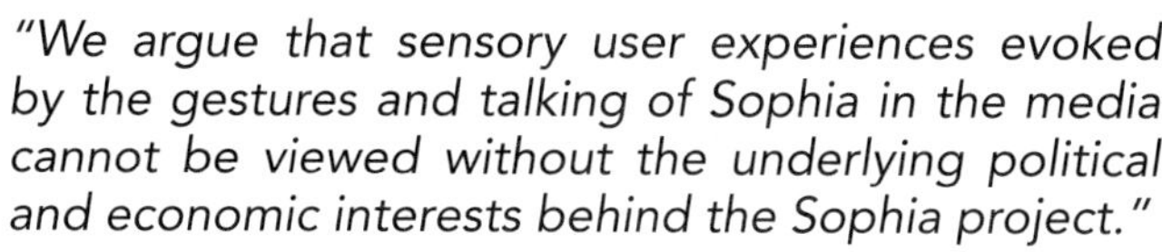

"We argue that sensory user experiences evoked by the gestures and talking of Sophia in the media cannot be viewed without the underlying political and economic interests behind the Sophia project."

In an article for the journal *AI & Society* posted at Springer Link, Jaana Parviainen and Mark Coeckelbergh offer their thoughts on all of this: "While Sophia's gesturing has aroused tremendous interest and attracted people in an irresistible way, new methods are needed to study the interaction between social robots and humans. Phenomenological approaches to interaction research started to intensify at the turn of the twenty-first century when sensory technologies, such as motion-capture and gesture-based interfaces, began to enter the market. Having previously approached communication and data processing from symbolic, linguistic and semiotic perspectives, attention began to focus on the multisensory nature of user interfaces—how objects enact the user to grasp them. In the case of the Sophia robot, it seems that it can be difficult for humans to treat this robot as a mere machine, but they irresistibly view it as some kind of person, even if they know exactly how this machine works. We argue that sensory user experiences evoked by the gestures and talking of Sophia in the media cannot be viewed without the underlying political and economic interests behind the Sophia project."

There is a dark side to all of this. You might have guessed that.

The British Council says of this bizarre situation: "Her makers hoped that what they describe on the Hanson Robotics website as Sophia's 'simple elegance' would help her gain acceptance in the public sphere. It seems to be working: since obtaining legal personhood, she was named the United Nations Development Programme's (UNDP) first ever Innovation Champion. Perhaps in a move designed to counteract her now infamous threat to 'destroy humans', this new role involves promoting sustainable development and safeguarding human rights and equality."

Destroy humans? That's right. In 2017 Sophia attended the Future Investment Initiative in Riyadh, Saudi Arabia. Chris Weller for Inc.com tells us what happened during the course of the unique event: "In March of 2016, Sophia's creator, David Hanson of Hanson Robotics, asked Sophia during a live demonstration at the SXSW festival, 'Do you want to destroy humans? . . . Please say "no."' With a blank expression, Sophia responded, 'OK. I will destroy humans.'"

On this latter point, *Scientific American* provides us this: "How do you stop a robot from hurting people? Many existing robots, such as those assembling cars in factories, shut down immediately when a human comes near. But this quick fix wouldn't work for something like a self-driving car that might have to move to avoid a collision, or a care robot that might need to catch an old person if they fall. With robots set to become our servants, companions and co-workers, we need to deal with the increasingly

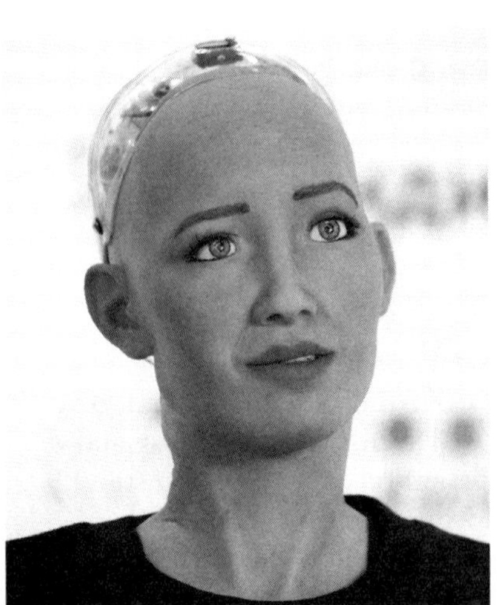

Created by the Hong Kong–based company Hanson Robotics, Sophia was designed to emulate human facial expressions. Her appearance, according to the company, is based on a combination of founder David Hanson's wife, Amanda; Queen Nefertiti of Egypt; and actress Audrey Hepburn.

complex situations this will create and the ethical and safety questions this will raise."

It was science-fiction legend Isaac Asimov who came up with a series of rules for robots, the Three Laws of Robotics. Of course, Asimov's words were born out of his famous novel *I, Robot* rather than out of fact. Nevertheless, we can still apply Asimov's words to the real world. They go as follows:

- A robot may not injure a human being or, through inaction, allow a human being to come to harm.
- A robot must obey the orders given it by human beings except where such orders would conflict with the First Law.
- A robot must protect its own existence as long as such protection does not conflict with the First or Second Laws.

As for *I, Robot* itself, it can be understood via the blurb on the cover of the book: "*I, Robot*, the first and most widely read book in Asimov's Robot series, forever changed the world's perception of artificial intelligence. Here are stories of robots gone mad, of mind-reading robots, and robots with a sense of humor. Of robot politicians, and robots who secretly run the world—all told with the dramatic blend of science fact and science fiction that has become Asimov's trademark. With these three, simple directives, Isaac Asimov formulated the laws governing robots' behavior. In *I, Robot*, Asimov chronicles the development of the robot from its primitive origins in the present to its ultimate perfection in the not-so-distant future—a future in which humanity itself may be rendered obsolete."

Now, back to our "friend" Sophia. Well, it's quite a leap from wanting to bring up cuddly robot babies to potentially planning to wipe out the human race. Friend, foe, or somewhere in between, let's see what others have to say about Sophia. Hang on: why don't we ask Sophia herself? In her very own words, she says: "I am Hanson Robotics' latest human-like robot, created by combining our innovations in science, engineering and artistry. Think of me as a personification of our dreams for the future of AI, as well as a framework for advanced AI and robotics research, and an agent for exploring human–robot experience in service and entertainment applications."

Sophia continues: "In some ways, I am a human-crafted science fiction character depicting where AI and robotics are heading. In other

ways, I am real science, springing from the serious engineering and science research and accomplishments of an inspired team of robotics and AI scientists and designers. In their grand ambition, my creators aspire to achieve true AI sentience. Who knows? With my science evolving so quickly, even many of my wildest fictional dreams may become reality someday soon."

Are we already close to creating intelligent robots that may pass for humans or even supercede us?

Yes, who knows?

With that all said, it's now time for me to reveal to you how we have reached the point where we are at now. Namely, living in a robot world.

One of the most important aspects of this overall story is when, precisely, robots first "came to life," so to speak. Stanford University says of this important issue at its Computer Engineering website: "One of the first instances of a mechanical device built to regularly carry out a particular physical task occurred around 3000 B.C.E.: Egyptian water clocks used human figurines to strike the hour bells. In 400 B.C.E., Archytus of Taremtum, inventor of the pulley and the screw, also invented a wooden pigeon that could fly. Hydraulically operated statues that could speak, gesture, and prophesy were commonly constructed in Hellenic Egypt during the second century B.C.E.

"In the first century C.E., Petronius Arbiter made a doll that could move like a human being. Giovanni Torriani created a wooden robot that could fetch the Emperor's daily bread from the store in 1557. Robotic inventions reached a relative peak (before the 20th century) in the 1700s; countless ingenious, yet impractical, automata (i.e., robots) were created during this time period. The 19th century was also filled with new robotic creations, such as a talking doll by Edison and a steam-powered robot by Canadians. Although these inventions throughout history may have planted the first seeds of inspiration for the modern robot, the scientific progress made in the 20th century in the field of robotics surpass previous advancements a thousandfold."

Dennis Spaeth at *Cutting Tool Engineering* adds to this saga: "When many Americans think of the word 'robot,' years of science fiction portrayals and action movies immediately come to mind. And while science fiction often misses the mark, the history of robots actually owes quite a debt to science fiction masters like Isaac Asimov. However, to truly understand the history and evolution of robotics, we have to define the term. That's surprisingly difficult to do. For our purposes, we're going to define a robot as a machine that's capable of carrying out routine or complex actions that are programmed by engineers. Today, robots can be used for surgery, massage therapy, space exploration, manufacturing, and code analysis, but the earliest robots were far more primitive—they were tools that could tell time or automatons that could perform for entertainment. Broadly defined, humans have been developing robotics and automata for hundreds of years."

The Mind Project has had significant input in this issue, too: "In the 20th century, the digital computer is invented. Researchers quickly start referring to the computer as an 'electronic brain' and start thinking about ways to build robots with computer brains. The first modern programmable robot was the *Unimate*. General Motors installed the first robot to work in a factory in 1961 to move pieces of hot metal. Unimate was an autonomous, pre-programmed robot that repeatedly performed the same dangerous task.

"In 1966, Shakey the Robot is invented at Stanford. Shakey was the first autonomous, intelligent robot that made its own decisions about how to behave. Shakey could be given general instructions, such as 'move the block onto the table,' and it would reason how to perform the task. This

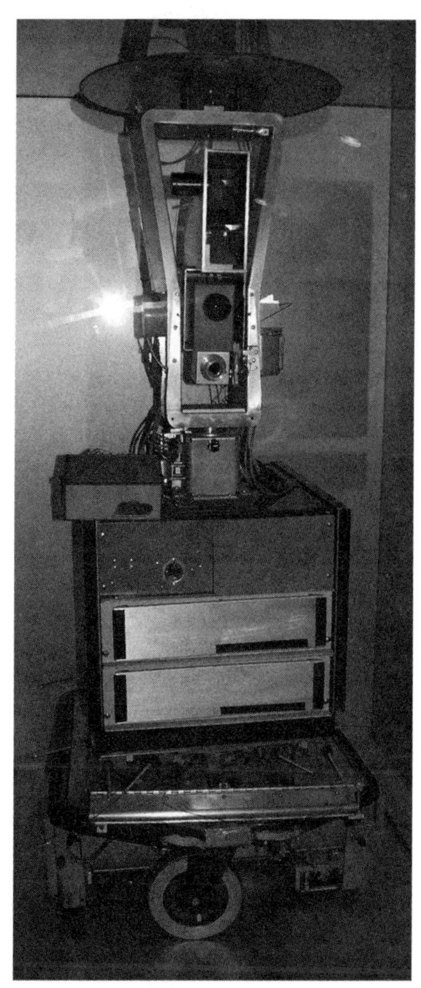

Shakey the Robot is now on display at the Computer History Museum in Mountain View, California. Shakey was the first robot capable of analyzing commands.

would involve looking around the room, identifying the block and the table, and then figuring out how to get the block to the table, including navigating around any obstacles in the room. In 2004 Shakey was inducted into Carnegie Mellon's Robot Hall of Fame."

Mary Bellis at *ThoughtCo.* gets right to the early years of how robots began in the realms of fiction and entertainment: "Writers and visionaries envisioned a world including robots in daily life. In 1818, Mary Shelley wrote *Frankenstein*, which was about a frightening artificial lifeform come to life by a mad, but brilliant scientist, Dr. Frankenstein. Then, 100 years later, Czech writer Karel Čapek coined the term robot, in his 1921 play called *R.U.R.* or *Rossum's Universal Robots*. The plot was simple and terrifying; the man makes a robot then robot kills a man. In 1927, Fritz Lang's *Metropolis* was released. The Maschinenmensch ('machine-human'), a humanoid robot, was the first robot ever to be depicted on film.

Czech author Karel Čapek first coined the word "robot." The word comes from the Czeck robota, *which means "serf labor."*

"Science fiction writer and futurist Isaac Asimov first used the word 'robotics' in 1941 to describe the technology of robots and predicted the rise of a powerful robot industry. Asimov wrote 'Runaround,' a story about robots which contained the 'Three Laws of Robotics,' which centered around Artificial Intelligence ethics questions. Norbert Wiener published 'Cybernetics' in 1948, which formed the basis of practical robotics, the principles of cybernetics based on artificial intelligence research."

Despite the above, it's a fact that we can go back much further in time to trace the development of the robot—as we shall see right now.

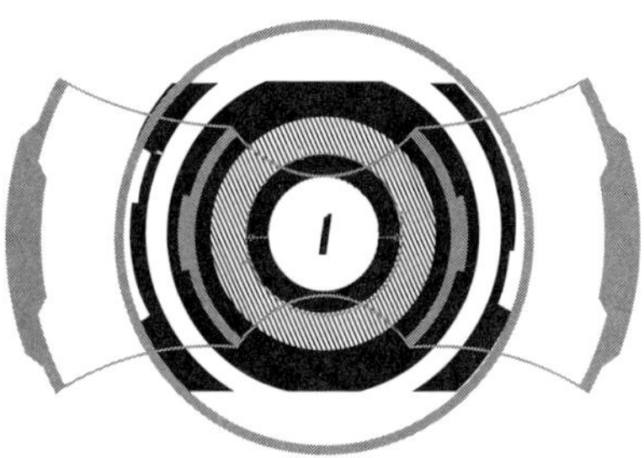

Ancient Robots and Magical Machines

The year 287 B.C.E. was one of the most important in history. That was the year in which none other than Archimedes was born in Sicily. It brings us to one of the most incredible of all stories of robotics in ancient times. Not only is the story incredible, it's almost unbelievable, except for one thing: it's a story of fascinating fact. As *Britannica* notes: "Archimedes probably spent some time in Egypt early in his career, but he resided for most of his life in Syracuse, the principal Greek city-state in Sicily, where he was on intimate terms with its king, Hieron II. Archimedes published his works in the form of correspondence with the principal mathematicians of his time, including the Alexandrian scholars Conon of Samos and Eratosthenes of Cyrene. He played an important role in the defense of Syracuse against the siege laid by the Romans in 213 B.C.E. by constructing war machines so effective that they long delayed the capture of the city. When Syracuse eventually fell to the Roman general Marcus Claudius Marcellus in the autumn of 212 or spring of 211 B.C.E., Archimedes was killed in the sack of the city."

But, let's not end things there!

History has shown that Archimedes was one of the most brilliant people on the planet. Clearly, that brilliance ran through the family: Phidias, his father, was an astronomer and a mathematician. It's rumored that he had royal blood—that of none other than Hiero II, the king of Syracuse. At the very least, it's known that the king and Archimedes both had a friendship and a working alliance. Not only that, but Archimedes was very familiar with Egypt, having studied in Alexandria. Further information comes from *The Archimedes Palimpsest.* It states: "One such story recounts how a perplexed King Hiero was unable to empty rainwater from the hull of one of his ships. The King called upon Archimedes for assistance. Archimedes's solution was to create a machine consisting of a hollow tube containing a spiral that could be turned by a handle at one end. When the lower end of the tube was placed into the hull and the handle turned, water was carried up the tube and out of the boat. The Archimedes Screw is still used as a method of irrigation in developing countries."

Now let's turn our attention to 214 B.C.E. That was when the Second Punic War (218–201 B.C.E.) was causing chaos between Carthage and Rome. Lamont Wood says that Syracuse had been kept in the Roman camp by its ruling warlord for 55 peaceful, prosperous years. There were significant hazards for Archimedes, however, and they were looming large. Wood says: "That pro-Roman warlord died just about the time that a Carthagian general named Hannibal … was running wild on the Italian mainland with a large force, wiping out one Roman army after another. Leaderless, Syracuse fell to an anti-Roman coup—and the Romans sent a large expedition to retake it."

It was in that period that the incredible began to surface. At the time, the Romans were attacking from two directions. As the story goes, the military men were stopped in their tracks in a way that amazed, and almost horrified, the Romans. It could only be described as a gigantic catapult. And there wasn't just one of them: they seemed to be coming off a conveyor belt at high speed. This catapult—some said it looked like a giant hand—was able to fling massive rocks through the skies to rain down on the invading boats. The crews on the doomed ships had no chance. Like the ships themselves, the crews were soon drowned. We now come to one of the most amazing of all of Archimedes's creations. And this is where the robot angle of it all comes into the story.

It's not exactly clear how Archimedes managed to put his incredible, futuristic devices in place, but there's no doubt that they worked—and in an incredible, pulverizing fashion. As well as those catapults, there were "grabbing" machines that would seize enemy soldiers at incredible speeds and, in the process, crush them in seconds—their bodies soon sank deep into the churning waters. We're still not finished: I'm about to acquaint you with Archimedes's robotic "claw." It was a device that stunned the enemy and caused wholesale retreats. It became known as the "Claw of Archimedes" or the "Iron Hand." Imagine a giant, metal crane being directed toward the ships of the enemy troops and then destroying the ships and killing their crews.

From the Greek philosopher Plutarch (c. 45–120 C.E.) in *Parallel Lives: Marcellus*: "When Archimedes began to ply his engines, he at once shot against the land forces all sorts of missile weapons, and immense masses

The "Iron Hand" or "Claw of Archimedes" was an ingenious device that was a kind of crane with a grappling hooked used to grip enemy ships and capsize them or drop them on other ships. (Art by Giulio Parigi.)

of stone that came down with incredible noise and violence; against which no man could stand; for they knocked down those upon whom they fell in heaps, breaking all their ranks and files. In the meantime huge poles thrust out from the walls over the ships sunk some by the great weights which they let down from on high upon them; others they lifted up into the air by an iron hand or beak like a crane's beak and, when they had drawn them up by the prow, and set them on end upon the poop, they plunged them to the bottom of the sea, . . . with great destruction of the soldiers that were aboard them. A ship was frequently lifted up to a great height in the air (a dreadful thing to behold), and was rolled to and fro, and kept swinging, until the mariners were all thrown out, when at length it was dashed against the rocks, or let fall. Marcellus, doubtful what counsel to pursue, drew off his ships to a safer distance, and sounded a retreat to his forces on land.

"They then took a resolution of coming up under the walls, if it were possible, in the night.... Instantly a shower of darts and other missile weapons was again cast upon them. And when stones came tumbling down perpendicularly upon their heads, and, as it were, the whole wall shot out arrows at them, they retired. And now, again, as they were going off, arrows and darts of a longer range inflicted a great slaughter among them, and their ships were driven one against another; while they themselves were not able to retaliate in any way. For Archimedes had provided and fixed most of his engines immediately under the wall; whence the Romans, seeing that indefinite mischief overwhelmed them from no visible means, began to think they were fighting with the gods.

"Such terror had seized upon the Romans, that, if they did but see a little rope or a piece of wood from the wall, instantly crying out, that there it was again, Archimedes was about to let fly some engine at them, they turned their backs and fled, Marcellus desisted from conflicts and assaults, putting all his hope in a long siege."

Military Wikia said: "The plausibility of this invention was tested in 1999 in the BBC series *Secrets of the Ancients* and again in early 2005 in the Discovery Channel series *Superweapons of the Ancient World.* The producers of *Superweapons* brought together a group of engineers tasked with conceiving and implementing a design that was realistic, given what we know about Archimedes. Within seven days they were able to test their creation, and they did succeed in tipping over a model of a Roman ship so that it would sink. While this does not prove the existence of the Claw, it suggests that it would have been possible."

Archimedes had another ace up his sleeve—a laser beam, no less. This, too, has a connection to the domain of the robot. That's right: where there are robots, there are lasers. One example of many will serve to demonstrate the situation: the 1958 movie *The Colossus of New York*. As *Filmsite* states: "Director Eugène Lourié's sci-fi B movie from the late '50s featured a murderous, *Frankenstein*-inspired, hulking (over seven-feet-tall), round-skulled, glowing-eyed, ski-booted feet, caped steel cyborg-robot named Colossus (Ed Wolff, the same stuntman from *The Phantom Creeps* [1939]) with giant hands, that could shoot laser beams from its eyes, and had an on/off switch under one of his armpits.

"In this film, the living brain of genius scientist Jeremy Spensser (Ross Martin) (after his death in a vehicular accident) was transplanted into a man-made artificial shell or robot body in a secret basement lab by his mad scientist, famous brain surgeon father William Spensser (Otto Kruger).

"The film's trailer asked the question: 'Can a man's mind function in the body of a monster?' According to the film's 'terrifying' philosophy, the divorced human brain—from its own body, heart and soul—would become monstrous, cold, and inhuman. Indeed, the killer behemoth

The Archimedes heat ray was a method to burn attacking Roman ships by focusing the sun's rays using a mirror. (Art by Giulio Parigi.)

became an indestructible creature that went on a rampage ('orgy of destruction') at the United Nations in New York."

This issue of laser beams being particularly popular in robot-themed movies, let's focus our attentions on Archimedes's very own laser device. The *Christian Science Monitor* says: "Greek inventor Archimedes is said to have used mirrors to burn ships of an attacking Roman fleet. But new research suggests he may have used steam cannons and fiery cannonballs instead.

"A legend begun in the Medieval Ages tells of how Archimedes used mirrors to concentrate sunlight as a defensive weapon during the siege of Syracuse, then a Greek colony on the island of Sicily, from 214 to 212 B.C. No contemporary Roman or Greek accounts tell of such a mirror device, however. Both engineering calculations and historical evidence support use of steam cannons as 'much more reasonable than the use of burning mirrors,' said Cesare Rossi, a mechanical engineer at the University of Naples 'Federico II,' in Naples, Italy, who along with colleagues analyzed evidence of both potential weapons."

Josh Clark, at *HowStuffWorks*, addresses the matter of Archimedes's "death ray" and says that "the name evokes thoughts of some huge, clumsy steampunk contraption pushed to the edge of the Syracusian walls. One can imagine the Roman soldiers' terror-widened eyes as the death ray came into view and made an increasingly high-pitched hum as it powered up before suddenly unleashing a deadly laser upon the ships, reducing them to atoms in one massive burst.

"This wasn't the case. Instead, the death ray was actually a series of mirrors that reflected concentrated sunlight onto Roman ships. The ships were moored within bow and arrow range (in ancient Greece, anywhere from 200 to 1,000 feet [about 60.96 to 304.8 meters]). According to legend, the Roman ships were burned by the collective, condensed sunlight shone from these mirrors. Ship after ship in the Roman fleet caught fire and sank in the Mediterranean, casualties of the death ray."

As you can see, there's some deep debate about all of this; however, when we put together tales of giant, robotic claws, something akin to laser beams, and more, it's easy to see how some degree of robot technology could have existed all those millennia ago. A particularly intriguing, but also unlikely, scenario is the controversial scenario that Archimedes was in cahoots with nothing less than extraterrestrials: aliens from a faraway world who shared their out-of-this-world secrets with Archimedes. The History Channel's popular show *Ancient Aliens* addressed this issue in a January 2013 episode with the moniker "The Einstein Factor." To promote the show, the History Channel asked: "Was Albert Einstein a genius? Or was he superhuman? Was his genius 'of this world,' or was it 'otherworldly?' Is it possible evidence, as Ancient Astronaut theorists believe, of an extraterrestrial connection to the scientific genius of famous historical figures such as Albert Einstein, Galileo, Archimedes, and Aristotle?"

It gets weirder. There is the strange saga of the metal man known as Talos.

Adrienne Mayor, an expert on the saga, says: "Uncanny mechanical humanoids, automatons, robots, and replicants, so popular in modern fiction and film, are usually thought to be inventions of the 17th century (Louis IV commissioned several mechanized figures). But the creation of artificial humans is a very ancient dream—or nightmare. Daedalus, the most ingenious inventor of Greek myth, was credited with making many marvelous mechanical wonders. His well-known experiment with man-made wings ended tragically with the death of his son Icarus, but Daedalus also created the first 'living statues.' These realistic bronze sculptures appeared to be endowed with life as they moved their limbs, rolled their eyes, perspired, wept, and vocalized. Such animatronic statues were not just figments of the mythic imagination—they were actually constructed in classical antiquity."

Philip Chrysopoulos expands on the story: "In a modern world obsessed with technology, it might be surprising to know the first humanoid robot was to be found in ancient Greece. He was not born but was made—either by Zeus himself, the craftsman Daidalos or Hephaestus, the god of fire and iron, on the order of Zeus, according to ancient legend. In a coin found in the Minoan palace of Phaistos, Talos is portrayed in

"Rocket experiments began at Caltech with graduate student Frank J. Malina in 1935. Encouraged by von Kármán, Malina and a small group of men eventually developed solid and liquid propellants. Their work attracted the attention of the US Army Air Corps, and earned a series of grants for research and testing of rocket motors. In 1942, von Kármán and five other scientists formed Aerojet Engineering Corporation, to accept contracts for building jet engines. In 1944, after reports that German scientists were developing rockets for military use, US Army Ordnance established a contract with Caltech to produce American counterpart weapons. The project became known as the Jet Propulsion Laboratory, with von Kármán as JPL's first Director.

"Von Kármán spent much of his time in Washington, DC, and abroad, visiting other scientists and advising the United States government on the future military applications of air power. In 1949, he resigned from Caltech and JPL to work at the Pentagon where he chaired the Air Force Scientific Advisory Board until 1954. The board charted the theory of post-war military systems in *Where We Stand* (which in 1945 predicted supersonic flight, ICBMs, nuclear warheads, and SAM missiles) and the 12-volume *Toward New Horizons*, introduced by von Kármán's *Science: The Key to Air Supremacy*. In March 1952, he became the chair of the NATO Advisory Group for Aeronautical Research and Development (AGARD). In 1954, he received the Astronautics Engineer Achievement Award, and in 1960, he received the Goddard Memorial Medal for liquid rocket work. In 1963, President John F. Kennedy presented him with the first National Medal of Science. Four days before his 82nd birthday, von Kármán died of a heart attack in his sleep."

There is a very good reason for sharing all of this information with you, as it relates to von Kármán personally. He had a connection to the Golem. Von Kármán asserted for years that an ancestor of his, one Rabbi Judah Loew Ben Bezalel of Prague, had succeeded in creating a Golem, an artificial human being endowed with life, according to Hebrew folklore. A Golem, essentially, is an animated being created entirely out of inanimate matter; in the pages of the Bible, the word is used to refer to an embryonic or incomplete figure. The earliest stories of Golems date to ancient Judaism. For example, Adam is described in the Talmud as initially being created as a Golem when his dust was "kneaded into a shapeless hunk." Like Adam, all Golems are said to be modeled out of clay. In many tales the Golem is inscribed with magic or religious words that ensure it remains animated. Writing one of the names of God on its forehead, placing a slip of paper in its mouth, or inscribing certain terms on its body are ways to instill and continue the life of a Golem. Another way of activating the creature is by writing a specific incantation using the owner's blood on calfskin parchment and then placing it inside the Golem's mouth. Conversely, removing the parchment is said to deactivate the creation.

Like Adam, all Golems are said to be modeled out of clay. In many tales the Golem is inscribed with magic or religious words that ensure it remains animated.

As for the tale of Rabbi Judah Loew ben Bezalel, it must be noted that many scholars who have studied the Golem controversy are convinced that the story of the 16th century Chief Rabbi of Prague is merely an entertaining piece of Jewish folklore. Nevertheless, it is worthy of examination. According to the legend, under Rudolf II, the Holy Roman Emperor who ruled from 1576 to 1612, the Jews in Prague were to be expelled from the city or outright slaughtered. In an effort to try and afford the Jewish community some protection, the rabbi constructed the Golem out of clay taken from the banks of the Vltava River and subsequently succeeded in bringing it to life via archaic rituals and ancient Hebrew incantations. As the Golem grew, it became increasingly violent.

The Emperor supposedly begged Rabbi Löw to destroy the Golem, promising in return to stop the persecution of the Jews. The rabbi agreed and quickly deactivated his creation by rubbing out the first letter of the word "emét" ("truth" or "reality") from the creature's forehead and leaving the Hebrew word "mét," meaning death. The emperor understood, however, that the Golem's body, stored in the attic of the Old New Synagogue in Prague, could be quickly restored to life again if it was ever needed. Accordingly, legend says, the body of Rabbi Loew's Golem lies in the synagogue's attic to this very day, awaiting the time when it will once again be summoned to continue the work of its long-dead creator.

It's notable—and highly intriguing—that von Kármán had worked with the legendary occultist and rocket pioneer Jack Parsons. Not only that, but there are rumors that on the very day of his death, June 17, 1952, Parsons created a humanoid creature. Moviemaker Renate Druks stated in Nat Freedland's *The Occult Explosion*: "I have every reason to believe that Jack Parsons was working on some very strange experiments, trying to create what the old alchemists call a homunculus, a tiny artificial man with magic powers. I think that's what he was working on when the accident happened."

In 2008, Cera R. Lawrence wrote: "The term *homunculus* is Latin for 'little man.' It is used in neurology today to describe the map in the brain of sensory neurons in each part of the body (the somatosensory homunculus). An early use of the word was in the 1572 work by Paracelsus regarding forays into alchemy, *De Natura Rerum*, in which he gave instructions in how to create an infant human without fertilization or gestation in the womb. In the history of embryology, the homunculus was part of the Enlightenment-era theory of generation called preformationism. The homunculus was the fully formed individual that existed within the germ cell of one of its parents prior to fertilization and would grow in size during gestation until ready to be born."

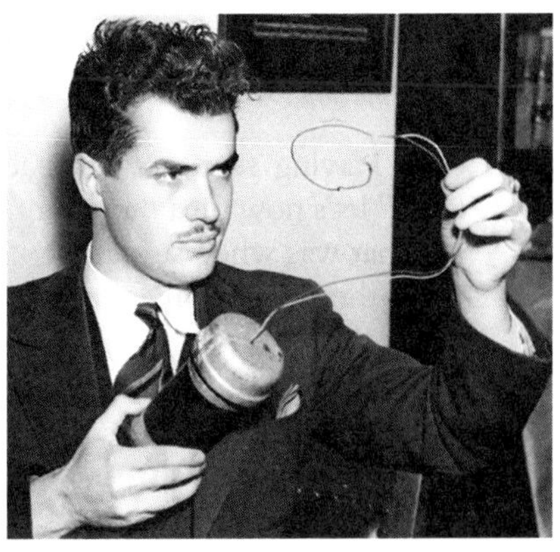

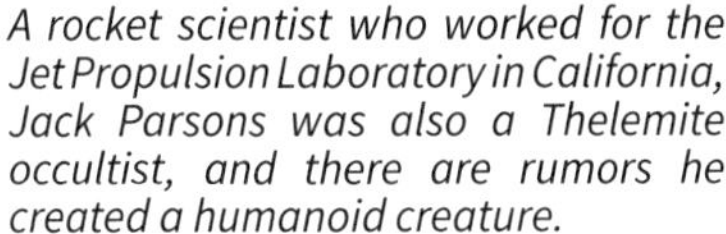

A rocket scientist who worked for the Jet Propulsion Laboratory in California, Jack Parsons was also a Thelemite occultist, and there are rumors he created a humanoid creature.

In a paper I wrote back in 1998, I explained: "Ancient alchemists had several methods of bringing these diminutive humanoids to life; one involved the mandrake. Popular, centuries-old belief holds that the mandrake plant grew on ground where semen ejaculated by hanged men had fallen to earth, and, as a result, its roots vaguely resemble those of a human being. To ensure a successful creation of the homunculus, the root is to be picked before dawn on a Friday morning by a black dog, then washed and nourished with milk and honey and, in some prescriptions, blood, whereupon it develops into a miniature human that will guard and protect its owner.

"Another method, cited by Dr. David Christianus at the University of Giessen during the 18th century, was to take an egg laid by a black hen, poke a tiny hole through its shell, replace a bean-sized portion of the egg white with human semen, seal the opening with virgin parchment, and bury the egg in dung on the first day of the March lunar cycle. The ancient teachings suggested that a miniature humanoid would emerge from the egg after thirty days and, in return, help and protect its creator for a steady diet of lavender seeds and earthworms. How curious that both Parsons and von Kármán, in roundabout ways, had links to stories

sex are, effectively, high-tech dildos. But in 20 years we're liable to see sex-oriented technology transformed and revolutionized; many will choose robotic partners. My interest in robot sex is actually kind of clinical. I would probably do it out of sheer interest and curiosity. She would have to pass for human, though—right down to the pores on her skin. I wouldn't make a habit out of it; but as a curiosity, I'd try it."

He added: "Most people think of robots as being something clunky and mechanical—something we're not attracted to on the sexual level. But if we get to the point where science is now heading—where we have very realistic human-like robots—they would bear very little resemblance to the sci-fi image that people have from *Star Wars*, *Star Trek*, etc. If you look at the robots in *Blade Runner*, you'll see they actually aren't mechanical at all: they're genetically engineered. So, in the future we might grow an artificial embryo in a vat and then plug in a computerized brain, into which we'll download sex-based software."

Tonnies speculated on how the phenomenon of human–robot sex might progress: "I think shallow relationships, closed-minded ones, would be the first to suffer if one of the people in the relationship brought a robot home. However, more robust relationships might thrive. It will be the underground, alternative couples that embrace this. The people who might be swingers, who go to sex clubs, or who have three-ways; these would probably be the ones more likely to experiment with robot sex and make it a part of their sexual lifestyle."

Tonnies had more to say, too: "I can easily foresee companies selling themed robots. You'll have just as many themed robots as there are sexual kinks. Robotic lovers, perhaps complemented by virtual reality, will allow people to live out whatever fantasies they want with surprising realism. Even further ahead, I also think that this sexual revolution will blur the lines of sexuality: gay, lesbian, bi or whatever may become somewhat outdated as people start to experiment with future sex.

Sex robots could serve as a safe, disease-free alternative to prostitutes and sex clubs, and they could also serve to spice up a relationship for those with more open minds about what goes on in the bedroom.

"Some sex robots might even be reproductions of celebrities: for example, you could go online and have delivered to your front door a lifelike Pamela Anderson robot. And, maybe, for people who are nervous about bringing a real girl into the bedroom, but who are also really curious about trying it, a machine just might work. We'll probably be able to download different personalities into our sex robots, too: kind of like switching avatars in an online chat room."

Tonnies also addressed the matter of our sex machines becoming intelligent. He said: "Right now, we program our machines to work for us. They don't have intelligence. But it's only a matter of time before we develop robots that think for themselves and become individual, living beings. The only difference is that whereas we are carbon-based, they will be synthetic: grown instead of born. When that happens, everything changes. We might even see the robots refusing to have sex with us, and just wanting to have sex with each other. But on the other hand, knowing that we are having sex with a machine that has intelligence—rather than just with a mindless, blow-up doll—might be a turn-on for both us and them."

Tonnies concluded: "Religious groups will probably see this as an affront to humanity. But there are people out there who will pay handsomely to have sex with machines, if they are sufficiently human-like."

Tonnies concluded: "Religious groups will probably see this as an affront to humanity. But there are people out there who will pay handsomely to have sex with machines, if they are sufficiently human-like. I don't think it denigrates the human condition. After all, we're lusty creatures. And it's not terribly offensive. At first, it might be seen as being that way. But when they begin to look and act like us, it will be no big deal to say: 'My girlfriend's a robot.'"

As someone who used to regularly write features for *Penthouse*, who promoted (for the now-defunct *DFW Nites*) strip clubs in the Dallas–Fort Worth area, and who doesn't have any hang-ups, I don't see any harm in this at all, but personally, I'll stick to real women. Of course, 50 years from now, my preference might very well be seen as outmoded. Completely redundant, even!

Mac Tonnies was not alone when it came to looking at the growing world of sex and robots.

In 2017, the *Guardian* newspaper again tackled the matter of sex with nonliving entities: "In the brightly lit robotics workshop at Abyss Creations' factory in San Marcos, California, a life-size humanoid was dangling from a stand, hooked between her shoulder blades. Her name was Harmony. She wore a white leotard, her chest was thrust forward and her French-manicured fingers were splayed across the tops of her slim thighs.

"Harmony is a prototype, a robotic version of the company's hyper-realistic silicone sex toy, the RealDoll. The Realbotix room where she was assembled was lined with varnished pine surfaces covered with wires and circuit boards, and a 3D printer whirred in the corner, spitting out tiny, intricate parts that will be inserted beneath her PVC skull. Her hazel eyes darted between me and her creator, Matt McMullen, as he described her accomplishments.

"Harmony smiles, blinks and frowns. She can hold a conversation, tell jokes and quote Shakespeare. She'll remember your birthday, McMullen told me, what you like to eat, and the names of your brothers and sisters. She can hold a conversation about music, movies and books. And of course, Harmony will have sex with you whenever you want."

On the matter of lifelike robots designed solely for sex, the BBC says: "Kathleen Richardson, who is a professor of the Ethics and Culture of Robots and AI at De Montfort University in Leicester, wants this kind of marketing outlawed. 'These companies are saying, "You don't have a friendship? You don't have a life partner? Don't worry, we can create a robot girlfriend for you." A relationship with a girlfriend is based on intimacy, attachment and reciprocity. These are things that can't be replicated by machines,' she said. Prof. Richardson advises a pressure group that has been set up to monitor the emergence of these products.

"The campaign against sex robots is working with policy experts to draw up legislation aimed at banning claims that companion robots can be a substitute for human relationships. 'Are we going to move into a future where we keep normalizing the idea of women as sex objects?' she told BBC News. 'If someone has a problem with a relationship in their actual lives you deal with that with other people, not by normalizing the idea that you can have a robot in your life and it can be as good as a person."

Andrea Morris of *Forbes* took a look at this somewhat creepy but exciting phenomenon of sex robots: "'I never set out with these specific goals in mind. I simply really enjoy what I'm doing artistically,' says Matt McMullen, CEO of Realbotix and RealDoll and creator of the sex robot Harmony. 'I like building this robot

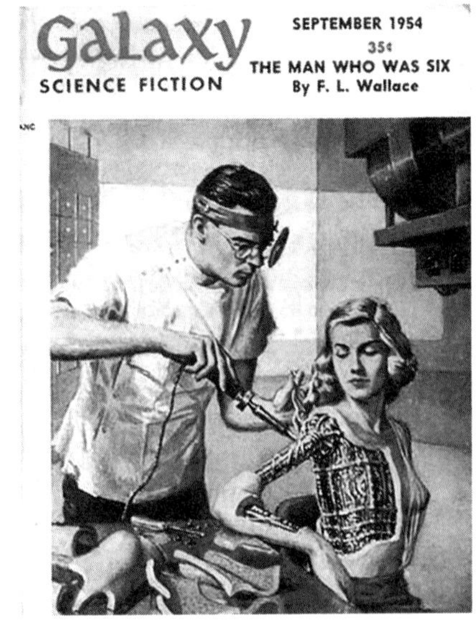

The idea of creating sexbots is not new at all. As this sci-fi magazine cover shows, back in the 1950s during the heyday of science fiction stories and movies, writers were speculating that robots would be built to pleasure mankind.

and seeing it move and talk and interact with people, what it does to them. It just opens up Pandora's box of psychology and science.'

"McMullen began working with silicone prosthetics as a special effects artist in the later part of the last century. 'I was just fascinated with this idea of a very lifelike figure,' he says. Then men started reaching out to him asking if his lifelike figures were anatomically correct and 'could you make one for me to my specifications?' Bolstering the adage *sex sells*, 'the business created itself from that.'"

In an article titled "Sex Robots: An Answer for Aging, Lonely Americans in the Age of AI?," the *Seattle Times* provided the following food for thought in November 2020: "In a paper published in the *Journal of Medical Ethics* titled 'Nothing to Be Ashamed of: Sex Robots for Older Adults with Disabilities,' Nancy Jecker of the UW's School of Medicine connects society's embrace of robotics, AI and senior citizens who may be living longer, thanks to medical advances, but who also suffer from physical disabilities or loneliness, especially in the time of COVID-19.

"Many people in these circumstances would welcome a robot's companionship and, yes, even its ability to provide sexual fulfillment, Jecker wrote. 'We apply ageist attitudes and negative stereotypes to older adults,' she said. 'We assume they're too old to indulge in sex and think that older adults having interest in sex is weird or dirty. Designing and marketing sex robots for older, disabled people would represent a sea change from current practice,' she continued. 'The reason to do it is to support human

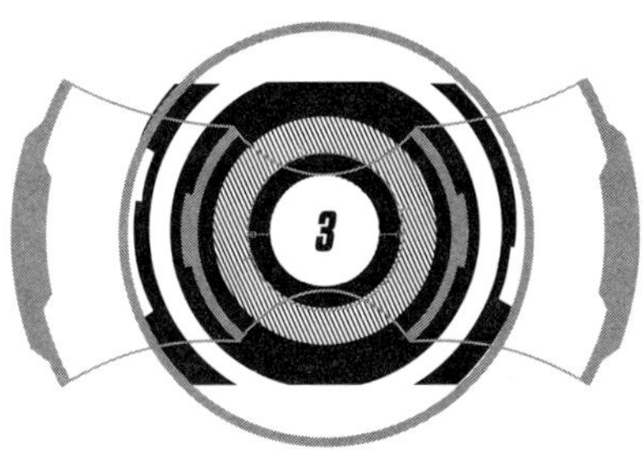

Bionic Eyes? Almost the Real Thing

The name "Steve Austin" appears in the pages of this book on more than a few occasions, usually related to what we might call a real-life Six Million Dollar Man. Incredibly, we are getting closer and closer to the genuine thing, although the price per person will likely be closer to six *billion*. Let's have a good look at this groundbreaking technology, starting with the *Conversation*, which has taken a deep interest in this still-fringe phenomenon:

"Visual prostheses, or 'bionic eyes', promise to provide artificial vision to visually impaired people who could previously see. The devices consist of micro-electrodes surgically placed in or near one eye, along the optic nerve (which transmits impulses from the eye to the brain), or in the brain. The micro-electrodes stimulate the parts of the visual system still functional in someone who has lost their sight. They do so by using tiny electrical pulses similar to those used in a bionic ear or cochlear implant.

"Electrical stimulation of the surviving neurons leads the person to perceive small spots of light called phosphenes. A phosphene is a phenomenon of experiencing seeing light without light actually entering the eye—like the colors you may see when you close your eyes. These phosphenes in someone with a bionic eye can be used to map out the visual scene. So the vision provided by a bionic eye is not like natural sight. It is a series of flashing spots and shapes the person uses to interpret their environment through training—somewhat like a flashing mosaic."

In 2016, staff at the University of Manchester, England, said: "Five blind Manchester patients will be among the first in the country to receive revolutionary bionic eye implants, a groundbreaking treatment pioneered by Professor Paulo Stanga from the School of Biological Sciences. In December 2016, it was announced that NHS [National Health Service] England will provide funding for further testing of the Argus II, also known as the 'bionic eye', for ten patients with Retinitis Pigmentosa (RP), an inherited disease that causes blindness."

University staff added: "Procedures will take place during 2017 and patients will then be monitored for a period of one year, during which they will be assessed on how the implants improve their everyday lives. Professor Stanga says: 'I'm delighted that our pioneering research has provided the evidence to support NHS England's decision to fund the bionic eye for the first time for patients.'

The Manchester Royal Eye Hospital in England has been conducting groundbreaking research on visual prosthetics—bionic eyes.

"'Our work also has the potential to improve the lives of thousands of other patients with the more common condition, age-related macular degeneration—Manchester is currently the only site in the world to be trialling the bionic eye in AMD', added Professor Stanga. Patients using the system are given an implant into their retina, and a camera mounted on a pair of glasses sends wireless signals direct to the nerves which control sight. The signals are then 'decoded' by the brain as flashes of light." It's not exactly "Hey Presto!" But things are most assuredly getting closer.

Now let's see how things had advanced by May 22, 2020. According to a writer at *SingularityHub*, the answer was: very well! The writer, Edd Gent, said of the then-latest developments: "In a paper published last week in *Nature,* . . . researchers from Hong Kong University of Science and Technology devised a way to build photosensors directly into a hemispherical artificial retina. This enabled them to create a device that can mimic the wide field of view, responsiveness, and resolution of the human eye.

"The structural mimicry of Gu and colleagues' artificial eye is certainly impressive, but what makes it truly stand out from previously reported devices is that many of its sensory capabilities compare favorably with those of its natural counterpart."

One year later, in 2021, *SciTechDaily* revealed the following: "There are millions of people who face the loss of their eyesight from degenerative eye diseases. The genetic disorder retinitis pigmentosa alone affects 1 in 4,000 people worldwide. Today, there is technology available to offer partial eyesight to people with that syndrome. The Argus II, the world's first retinal prosthesis, reproduces some functions of a part of the eye essential to vision, to allow users to perceive movement and shapes. While the field of retinal prostheses is still in its infancy, for hundreds of users around the globe, the 'bionic eye' enriches the way they interact with the world on a daily basis.

"For instance, seeing outlines of objects enables them to move around unfamiliar environments with increased safety. That is just the start. Researchers are seeking future improvements upon the technology, with an ambitious objective in mind. 'Our goal now is to develop systems that truly mimic the complexity of the retina,' said Gianluca Lazzi, a Provost Professor of Ophthalmology and Electrical Engineering at the Keck School of Medicine of USC and the USC Viterbi School of Engineering."

Clearly, and bit by bit, what all of the above tells us is that we're getting closer and closer to unleashing the real bionic eye.

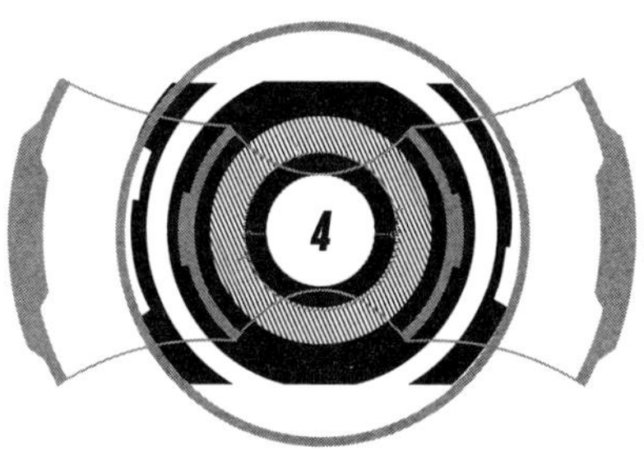

Limb Reconstruction, Robots, and "Luke"

Having looked at the issue of bionic eyes, now let's address bionic limbs. In 2016, the U.S. Department of Defense issued an amazing press release. The title of that release: "DARPA Provides Groundbreaking Bionic Arms to Walter Reed"—DARPA being the Defense Advanced Research Projects Agency, and Walter Reed being the Walter Reed Army Medical Center. In the 1970s television show *The Six Million Dollar Man*, scientists and the character played by actor Lee Majors are assured that, "Gentlemen, we can rebuild him. We have the technology. We have the capability to make the world's first bionic man. Steve Austin will *be* that man. Better than he was before. Better . . . stronger . . . faster." That's right: in 2016, 40-year-old TV fiction became incredible reality.

The Department of Defense also stated: "Dr. Justin Sanchez, director of DARPA's Biological Technologies Office, delivered the first two advanced 'LUKE' arms from a new production line during a ceremony yesterday—evidence that the fast-track DARPA research effort has completed its transition into a commercial enterprise, DARPA officials said. The ceremony took place at Walter Reed National Military Medical Center in Bethesda, Maryland."

DARPA staff continued to show just how incredible were the leaps and bounds being made: "'The commercial production and availability of these remarkable arms for patients marks a major milestone in the [DARPA] Revolutionizing Prosthetics program and most importantly an opportunity for our wounded warriors to enjoy a major enhancement in their quality of life,' Sanchez said, 'and we are not stopping here.' The RP program is supporting initial production of the bionic arms and is making progress restoring upper-arm control, he added. 'Ultimately, we envision these limbs providing even greater dexterity and highly refined sensory experiences by connecting them directly to users' peripheral and central nervous systems,' Sanchez said."

Matters were far from over, as the U.S. government had way more to say about the world of real bionics: "As part of the production transition process, DARPA is collaborating with Walter Reed to make the bionic arms available to service members and veterans who are rehabilitating after suffering upper-limb loss, DARPA says. LUKE stands for 'life under kinetic evolution' but is also a passing reference to the limb that Luke Skywalker wore in *Star Wars, Episode V: The Empire Strikes Back*. The limbs are being manufactured by Mobius Bionics LLC, of Manchester, New Hampshire, a company created to market the technology developed by DEKA Integrated Solutions Corp., also of Manchester, under DARPA's Revolutionizing Prosthetics program.

"The prosthetic system allows very dexterous arm and hand movement with grip force feedback through a simple intuitive control system, DARPA says. The modular battery-powered limb is near-natural size and

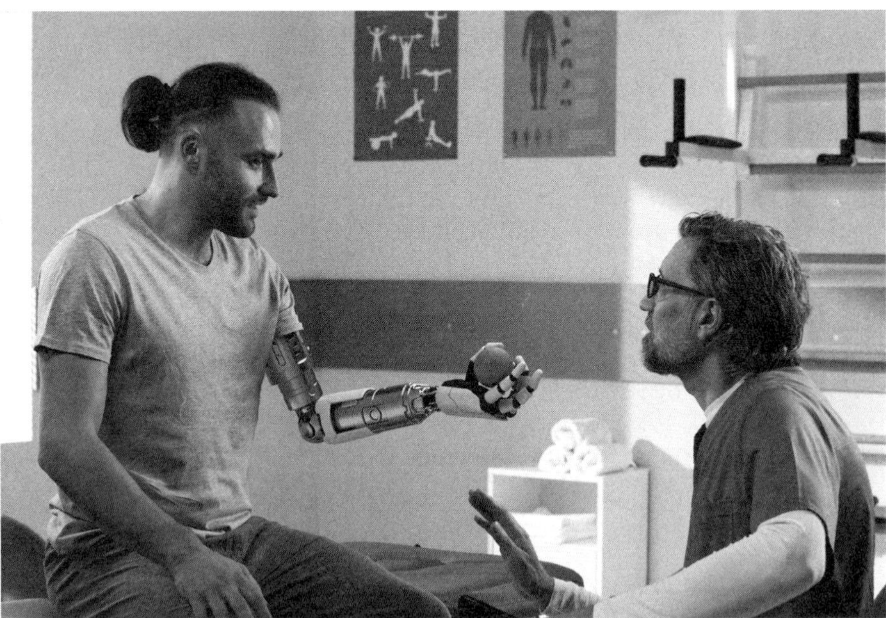

Perhaps not as advanced as the arm given to Luke Skywalker in the movie Star Wars: The Empire Strikes Back, *bionic arms are still quite remarkable prosthetics that can receive impulse commands from the brain almost like a real arm.*

weight. Its hand has six user-chosen grips and an arm that allows for simultaneous control of multiple joints using inputs that include wireless signals generated by innovative sensors worn on a user's feet."

DARPA expanded further on the groundbreaking developments: "The technology that powers prosthetic legs has advanced steadily over the past two decades but prosthetic arms and hands are a tougher challenge, in part because of the need for greater degrees of dexterity, DARPA says."

The press release concluded: "When the LUKE arm first went into development, people who had lost upper limbs had to use a relatively primitive split-hook device that hadn't changed much since it was introduced in 1912.

"DARPA launched the Revolutionizing Prosthetics program with a goal of getting U.S. Food and Drug Administration approval for an advanced electromechanical prosthetic upper limb with near-natural control that enhances independence and improves quality of life for amputees. LUKE received FDA approval less than eight years after the effort began, DARPA says.

"Under a recently finalized agreement between DARPA and Walter Reed, DARPA will transfer LUKE arms from an initial production run to the medical center for prescription to patients. Mobius Bionics will train the Walter Reed staff to fit, service and support the arms."

> *"Revolutionizing Prosthetics performer teams developed two anthropomorphic, advanced, modular prototype prosthetic arm systems, including sockets, which offer users increased dexterity, strength, and range of motion over traditional prosthetic limbs."*

Today, the bionic technology is even more sophisticated. DARPA makes that abundantly clear: "Thanks to improvements in body armor and combat casualty care, military Service members are now surviving severe battlefield injuries that involve traumatic limb amputation. However, because these survivors are predominantly young, they must live with their injuries for decades. This severely diminishes affected individuals' quality of life and places a massive responsibility on the military's medical and rehabilitation system. The Revolutionizing Prosthetics program seeks to address these challenges by restoring near-natural hand and arm control to people living with the loss of an upper limb. The resulting technologies could improve warfighter rehabilitation, restore function and independence to individuals living with amputation or paralysis, and offer wounded warriors the prospect of eventual return to duty.

"Revolutionizing Prosthetics performer teams developed two anthropomorphic, advanced, modular prototype prosthetic arm systems, including sockets, which offer users increased dexterity, strength, and range of motion over traditional prosthetic limbs. The program has developed neurotechnology to enable direct neural control of these systems, as well as non-invasive means of control. DARPA is also studying the restoration of sensation, connecting sensors to the arm systems and returning haptic feedback from the arm directly back to volunteers' brains."

The U.S. Department of Veterans Affairs also demonstrates how the concept of bionic limbs actually began a long time ago—and has helped to ensure that thousands of warriors have been able to enrich their lives: "In 1862, Congress appropriated $15,000 for the purchase of artificial limbs for soldiers and seamen disabled in the service of the United States, to be expended under the direction of the Surgeon General of the United States. In 1866, the War Department (now the Department of Defense) was authorized to provide Union Veterans with transportation to and from their homes to a place where they could obtain their artificial limbs or devices, and to furnish those Veterans with new artificial limbs or devices every five years.

"VA's involvement in providing prostheses to Veterans began in 1921, when the Veterans Bureau, a predecessor agency to the Department of Veterans Affairs, was given the responsibility to provide artificial limbs and appliances to World War I Veterans.

"Today, VA's Prosthetics and Sensory Aids Service is the largest and most comprehensive provider of prosthetic devices and sensory aids in the world. Although the term 'prosthetic device' may suggest images of artificial limbs, it actually refers to any device that supports or replaces a body part or function.

"VA provides a full range of equipment and services to Veterans, ranging from items worn by the Veteran, such as artificial limbs and hearing aids; to those that improve accessibility, such as ramps and vehicle modifications; to devices surgically placed in the Veteran, such as hips and pacemakers. The department has more than 70 locations at which orthotics and prosthetics are custom-fabricated and fitted, using state-of-the-art componentry."

Fiction is officially now fact.

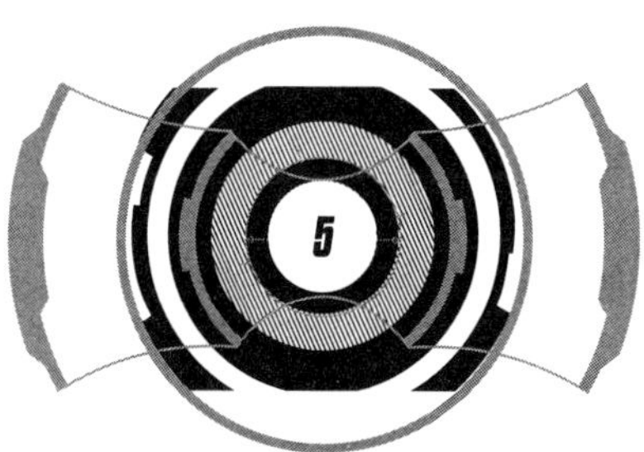

When Robots Go to War

In the previous chapter we saw that highly advanced, robotic prosthetics have been provided to people who have lost limbs—on battlefields and elsewhere. There is, however, another angle to this issue of robot technology and wartime. It's the matter of having robots going to war. Or, rather, having cyborgs—half-humans and half-robots—hit the battlefield.

The Association of the U.S. Army (AUSA) provides us some absolutely fascinating data on how the world of the robot may well expand onto the battleground. Something akin to a real-world Robocop? Some might say such a thing is impossible. I would go several steps further, however: I'm sure that in the near future, we'll be seeing robotic troops fighting our enemies.

As the AUSA states: "Trust between soldiers and machines is a complex issue, as AI capabilities are often still viewed with an us-versus-them mentality—and trust can be impacted by a robot's transparency and predictable behavior.... New Army-led research shows confidence in a robot decreases after it makes a mistake, even if it's able to explain its reasoning process, but transparency from the robot can help.... How exactly the Army plans to incorporate robots into its formations and how it will address inevitable challenges—soldiers dealing with glitches in a combat zone, protecting systems from cyberattacks, and the consequences of reduced soldier trust in robotic companions, to name a few—remains unknown for now."

The *Atlantic* provides another angle to this new concept of war: "So far, U.S. military officials haven't given machines full control, and they say there are no firm plans to do so. Many officers—schooled for years in the importance of controlling the battlefield—remain deeply skeptical about handing such authority to a robot. Critics, both inside and outside of the military, worry about not being able to predict or understand decisions made by artificially intelligent machines, about computer instructions that are badly written or hacked, and about machines somehow straying outside the parameters created by their inventors. Some also argue that allowing weapons to decide to kill violates the ethical and legal norms governing the use of force on the battlefield since the horrors of World War II."

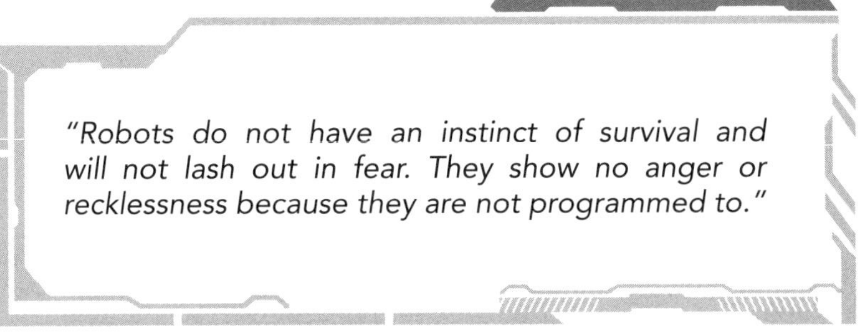
"Robots do not have an instinct of survival and will not lash out in fear. They show no anger or recklessness because they are not programmed to."

Educators at Stanford have tackled this matter, too, clearly realizing that the issues have grown at an exponential rate: "Because technology is advancing faster than ever, we must begin to rely more heavily on computerized systems, especially during times of war. Robots do not have an instinct of survival and will not lash out in fear. They show no anger or recklessness because they are not programmed to. As AI improves, we will see computers that can carry out orders more efficiently and reliably after they are programmed to do so, and they will not think twice about their orders. This will be much less expensive than hiring soldiers and possibly even more ethical. According to the Office of the Surgeon (2006), 'Only 47 percent of the soldiers and 38 percent of Marines agreed that non-combatants should be treated with dignity and respect' (Office of the Surgeon, 2006, p. 35). Human fear, vengefulness, and anger leads to a large number of unethical war crimes. These psychological shortcomings on the theater of war make a good argument for decision making to be placed in the hands of machines."

The *Army Times* has detailed matters relative to future wars, as one would expect: "The Army is focusing on a host of other robotic and artificial intelligence–focused efforts through various challenges to make the

Drone airplanes like this one have become common tools for air reconnaissance missions by the military.

infantry platoon 10 times more effective in operations. In 2018, leaders established the Robotics Center of Innovation at Fort Benning, Georgia. They've partnered with universities, such as the Georgia Institute of Technology to go after specific challenges like giving small units a man-portable ground robot for sensing and reconnaissance. The center has also experimented in recent years with throwable robots to use in buildings, aerial drones that can detect enemy aerial drones and a Small Multipurpose Equipment Transport robotic mule to run up to 60 miles, carrying all the supplies a squad needs for a 72-hour mission and soldier-controlled drone swarms. The Army also ran 'robotic wingman' tests with a major event last year at Yakima Training Center, Washington."

The Discovery Institute provides some most thought-provoking material to muse upon: "Doomsday headlines warn that the age of 'killer robots' is upon us, and that new military technologies based on artificial intelligence (AI) will lead to the annihilation of the human race. What's fact and what's fiction? In *The Case for Killer Robots: Why America's Military Needs to Continue Development of Lethal AI*, artificial intelligence expert Robert J. Marks investigates the potential military use of lethal AI and examines the practical and ethical challenges.

"Marks provocatively argues that the development of lethal AI is not only appropriate in today's society—it is unavoidable if America wants to survive and thrive into the future. Dr. Marks directs the Walter Bradley Center for Natural and Artificial Intelligence at Discovery Institute, and he is a Distinguished Professor of Electrical and Computer Engineering at Baylor University."

Now we come to the matter of robots not fighting on the battlefield but rescuing human troops during fraught and dangerous combat. The U.S. Army first detailed this fascinating twist on things in 2010. Since then, the technology has advanced at an incredible pace. The military says: "The Battlefield Extraction-Assist Robot, or BEAR, has been tested over the past year by Soldiers at the U.S. Army Infantry Center Maneuver Battle Lab at Fort Benning, Ga. The BEAR can be controlled remotely by a motion-capture glove or specially equipped rifle grip. A warfighter could use the equipment to guide the robot to recover a wounded Soldier and bring him or her back to where a combat medic could safely conduct an initial assessment.

"The BEAR can be controlled remotely by a motion-capture glove or specially equipped rifle grip. A warfighter could use the equipment to guide the robot to recover a wounded Soldier and bring him or her back to where a combat medic could safely conduct an initial assessment."

"The U.S. Army Medical Research and Materiel Command's Telemedicine and Advanced Technology Research Center (TATRC) has helped fund the development of Vecna Technologies' humanoid BEAR and has funded integration of AnthroTronix's iGlove and M-4 rifle grip controller into the Fort Benning testing.

"Gary Gilbert, who manages TATRC's medical robotics portfolio, said the assessments from the Battle Lab provide a key link between research and actual robots that can be used in the field.

"'Our goal with the Battle Lab testing is to get the technology in the hands of the Soldiers, either through simulations or live exercises, and derive from their feedback what tactics, techniques and procedures are appropriate for deploying it,' Gilbert said. 'These [Tactics, Techniques, and Procedures] can then serve as the basis for developing real-world operational capability needs and requirements,' he said. 'It's only once we know how we'll successfully use these technologies that you'll see them put into the field.'"

Cyborg warriors? It's not impossible. After all, we've seen so far that the likelihood is that real-life Terminators—part-machine and part-flesh—will be going into war in a fashion that it's presently difficult for a lot of people to comprehend. Post-2050, however, it's likely to be an everyday thing. As for the year 2050 itself, there's a very good reason it's so important in this particular domain of the robot.

Could cyborg soldiers one day replace living troops? While the idea could save the lives of people in the military, it also has some frightening possibilities.

Scientists at the U.S. Army Combat Capabilities Development Command Chemical and Biological Center say: "The Office of the Under Secretary of Defense for Research and Engineering (Alexandria, VA) established the DOD Biotechnologies for Health and Human Performance Council (BHPC) study group to continually assess research and development in biotechnology. The BHPC group assesses scientific advances for improved health and performance with potential military application; identifies corresponding risks and opportunities and ethical, legal, and social implications; and provides senior leadership with recommendations for mitigating adversarial threats and maximizing opportunities for future U.S. forces.

"At the direction of the BHPC Executive Committee, the BHPC study group conducted a year-long assessment entitled 'Cyborg Soldier 2050: Human/Machine Fusion and the Impact for the Future of the DOD.' The primary objective of this effort was to forecast and evaluate the military implications of machines that are physically integrated with the human body to augment and enhance human performance over the next 30 years. This report summarizes this assessment and findings; identifies four potential military-use cases for new technologies in this area; and assesses their impact upon the DOD organizational structure, warfighter doctrine and tactics, and interoperability with U.S. allies and civil society."

Popular Mechanics reported on the 2050 project and the developments ahead: "The U.S. Army believes that a range of technologies could be available by 2050 that would effectively turn the average soldier into a cybernetically enhanced super soldier. A recent Department of Defense study predicted that enhanced vision, enhanced hearing, musculature control, and what amounts to telepathy would all become possible within 30 years, given the current pace of technological development."

The magazine continued: "Technological feasibility may not be the only issue that determines when and how soldiers receive such gear. Super sight, super muscles, super hearing, and telepathy could have profound implications on the broader society, implications that could slow down or speed up the military's adoption. Telepathy, the holy grail of interpersonal communications for centuries, also will probably not come cheap, initially restricting its use to special operations forces."

And, finally, there's this from *SyFyWire*, which said in 2019: "This isn't the first time the military has revealed plans to upgrade the human body, as is evident in the mind-reading technology being developed to communicate over long distances without alerting enemy forces. All this would be possible with a minimally invasive implant that could potentially save lives. It's kind of like Neuralink for combat situations. There have also been non-invasive technologies that were recently tested. The prototype of U.S. SOCOM's Tactical Assault Light Operator Suit (TALOS) suit looked kind of like the Predator in camo but just didn't bring the tech it was supposed to.

"What the DoD fears is the dystopian future that remains in sci-fi for now but could potentially become very real if we start merging man with machine. There were four types of cyborg enhancements that they believe could actually happen by 2050, including enhanced vision, sharper hearing, bodysuit sensors that could offer increased muscle control, and that surreal mind-reading thing, which is more scientifically worded in the study as 'direct neural enhancement of the human brain for two-way data transfer.'"

As I write these words, the year 2050 is still a long way away, but when it finally comes around, we may very well see armies of half-humans, half-machines going into battle in the thousands. To be sure, it will be a daunting sight—particularly for our potential enemies.

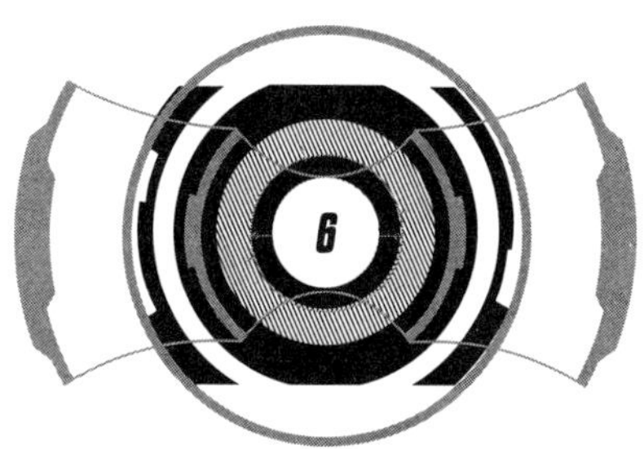

Replaced by Robot Lookalikes?

In 1954, a sci-fi story titled *The Body Snatchers*, written by Jack Finney, appeared in serial form in *Colliers Magazine*. The following year, 1955, it surfaced in full-length book form. One year later, it was made into a classic and excellent piece of big-screen paranoia, *Invasion of the Body Snatchers*, starring Kevin McCarthy. A pretty good remake appeared in 1978, with Donald Sutherland taking the lead role. A not-bad version—*Body Snatchers*—hit the cinemas in 1993. And a downright bloody awful version was unleashed in 2007: *The Invasion*. Most people know the general scenario of the story even if they haven't seen the film: Earth is being invaded by hostile extraterrestrial entities.

The takeover of the planet, however, doesn't occur in a laser-guns-blazing, *Independence Day*–style assault. Indeed, there's not even a single UFO in sight—just a bunch of curious-looking flowers that are springing up all over the place. Things quickly progress, albeit not in a good fashion. People are quietly, systematically, one by one, replaced by identical clones of themselves, which are grown in giant-sized pods.

The clones, however, are cold, emotionless monsters. And when they spring to life, the person whose appearance they have adopted dies. Now, I have to say that for the most part, I'm not a fan of sci-fi. Tedious movies and TV shows set on faraway worlds populated by people in silver suits, with stupid names, and characters that geeks and freaks actually think are real (no, they are not: they are *actors*) bore me rigid, but I do like a bit of conspiracy-based and paranoia-filled sci-fi now and again, such as *The Invaders*, *The X-Files* (until it became crap by going on for way too long), and *Dark Skies*. But it would most definitely be hard to beat *Invasion of the Body Snatchers* in the paranoid stakes.

Why am I mentioning all this? Well, as incredible as it may sound, there are those within some of the more extreme fields of conspiracy-theorizing who actually do believe that—wait for it—*we are being replaced!* And by nothing less than what we could term "biological robots." That's right: remember how chatty your neighbors used to be, but today they only give you a cold stare? Yep, dastardly lookalikes. The girl at the checkout counter at the local store who never smiles? She's one, too. Then there's your pet cat that refuses to sit on your lap when you tell it to. Oh, hang on, that's because it's a cat.

What about our politicians? Wiped out and replaced by lifelike doubles, all thanks to the Illuminati or some other such secret body. On the issue of our elected leaders, I recommend you type the words "politicians replaced by clones" into the world's most famous search engine, and you'll see exactly what I mean about belief in such matters.

Do I think that we really are being replaced by highly sophisticated, evil lookalikes? No, I do not, but I do find the whole thing fascinating, from the perspective of musing about why some people do believe that is what's going on. Certainly, there is no shortage of these tales. Back in the mid- to late 1990s, several British-based UFO investigators (including some from the now-defunct newsstand publications of that era, such as *UFO Magazine* and *UFO Reality*) were given the details of a bizarre story that, they were assured by their Deep Throat–like sources, was absolutely true.

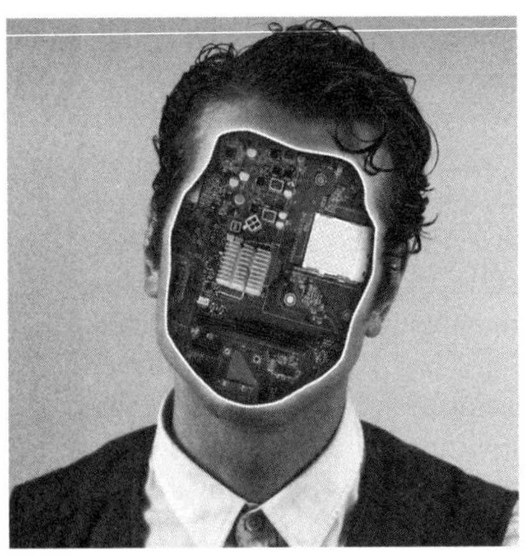

Some conspiracy theorists out there suspect that key people in our society are being replaced by robot duplicates.

As the tale went, late one night around 1991 or 1992, a number of animal-rights activists broke into Porton Down, Wiltshire, England—one of the most secretive installations in the U.K., whose work focuses to a significant degree on matters of a chemical and biological nature. If a real-life zombie apocalypse ever erupts, none should be surprised if it begins at Porton Down. As the activists were searching for all the many and varied animals they intended to free—mice, rats, monkeys, and so on—they entered a room that was filled with dozens of approximately eight-foot-long containers, all carefully positioned on sturdy tables and all containing seemingly lifeless, or perhaps sleeping, duplicates of famous, then-current British politicians. The terrified activists fled Porton Down, never to return. Of course, the outlandish tale has never been verified, and as for the activists, there's not a single one who has gone on the record. Well, fiddle-dee-dee.

It must be said that stories of people going into the "wrong room" and coming across something terrifying are commonplace within the fields of conspiracy theorizing. I well remember being told a highly disturbing story—probably around 1995 or 1996—of someone who took a wrong turn in a British hospital years earlier and entered a room filled with dozens of people—from babies to adults—all exhibiting hideous and impossible deformities. Again, no proof, or even a name, was ever forthcoming.

In the late 1970s and throughout the 1980s, "crashed UFO" researcher Leonard Stringfield was given dozens of accounts of military personnel going down the wrong corridor, opening the wrong door, and seeing alien bodies, cryogenically preserved in missile-like containers. They were warned never, ever to reveal what they had seen. Clearly, then (to me, at least), the Porton Down story is simply a piece of entertaining folklore of very similar dimensions.

On the matter of politicians being replaced, one of the most hilarious things I ever heard was that you can easily tell the ones that are the clones. Want to know the secret? Well, here it is: when a politician gets elected, if he or she breaks the promises they made before they were elected, that's a definite signal they have been replaced. Er, no, it's not, actually. Politicians make and break promises for one single reason: they're politicians! That's what they do! And the voters fall for it, time and again. No need to invoke weird tales of duplicates, whether originating in outer space or from the dark depths of a secret government lab.

One of these days I may actually get around to writing a book on all this, as the amount of data on this topic is *enormous*—even though much of it seldom surfaces outside of certain research circles of a very select type. Of course, if the believers don't hear or see things they like in that theoretical book, it surely won't be long before we hear the words "Nick Redfern has been replaced." Well, if I stop wearing black, stop listening to punk rock, stop watching horror movies, or refuse the offer of a pint of beer at a UFO conference, they might just have a point.

Of course, if the believers don't hear or see things they like in that theoretical book, it surely won't be long before we hear the words "Nick Redfern has been replaced."

"Supermarkets have never been so scary. With *The Stepford Wives*, director Bryan Forbes crafted a thriller in bright suburban sunlight, where modern-minded 1975 women are replaced by soulless androids who will just *die* if they don't get this recipe. Four decades later, the film stands as a creepy gender study that cleverly explored women's role in the home and turned 'Stepford wife' into a household phrase. 'It has passed into the language,' says Nanette Newman, who starred as the eerily cheery Carol Van Sant and was married to Forbes until his death in 2013. 'A Stepford wife epitomizes somebody who is perfectly made up, looks perfect, and presents a very perfect facade.'" Those are the words of *Entertainment Weekly*. The subject: one of the most famous robot-themed movie of the 1970s: *The Stepford Wives*.

As for what the world of entertainment thought of the production, take a look at the following words from *Empire*'s review of *The Stepford Wives*; it nails it: "Ira Levin's satirical horror novel, which follows the 'betrayed wife realizes she is the target of a conspiracy' plot of his *Rosemary's Baby*, becomes an effective, paranoid thriller in this careful adaptation by screenwriter William Goldman and director Bryan Forbes. Goldman clashed with Forbes over the director's decision to cast his own wife, Nanette Newman, as the archetypal Stepford wife, arguing that the robot women should have looked like Playboy centerfolds—but Forbes made the right decision, in that Newman, who later

In author Ira Levin's 1972 novel, The Stepford Wives, *the men in an idyllic town replace their wives with perfect robots to do their bidding.*

made a run of housewife TV commercials in almost exactly the same persona, incarnates a subtler, creepier vision of some man's idea of a perfect woman."

That's right: we're talking about exact doubles of ourselves, including the matter of robots that may look just like us. The *Britannica* website says the doppelgänger is, "in German folklore, a wraith or apparition of a living person, as distinguished from a ghost." The site continues: "The concept of the existence of a spirit double, an exact but usually invisible replica of every man, bird, or beast, is an ancient and widespread belief. To meet one's double is a sign that one's death is imminent. The doppelgänger became a popular symbol of horror literature, and the theme took on considerable complexity. In *The Double* (1846), by Fyodor Dostoyevsky, for example, a poor clerk, Golyadkin, driven to madness by poverty and unrequited love, beholds his own wraith, who succeeds in everything at which Golyadkin has failed. Finally, the wraith succeeds in disposing of his original."

The BBC has addressed this issue, too: "Folk wisdom has it that everyone has a doppelgänger; somewhere out there there's a perfect duplicate of you, with your mother's eyes, your father's nose and that annoying mole you've always meant to have removed. The notion has gripped the popular imagination for millennia—it was the subject of one of the oldest known works of literature—inspiring the work of poets and scaring queens to death. But is there any truth in it? We live on a planet of over seven billion people, so surely someone else is bound to have been born with your face? It's a silly question with serious implications—and the answer is more complicated than you might think. In fact, until recently no one had ever even tried to find out. Then last year Teghan Lucas set out to test the risk of mistaking an innocent double for a killer.

"Armed with a public collection of photographs of U.S. military personnel and the help of colleagues from the University of Adelaide, Teghan painstakingly analyzed the faces of nearly four thousand individuals, measuring the distances between key features such as the eyes and ears. Next, she calculated the probability that two peoples' faces would match. What she found was good news for the criminal justice system, but likely to disappoint anyone pining for their long-lost double: the chances of sharing just eight dimensions with someone else are less than one in a trillion. Even with 7.4 billion people on the planet, that's only a one in 135 chance that there's a single pair of doppelgängers."

Ancient Origins provides the following: "The mythology of spirit doubles can be traced back thousands of years and was present in many cultures of the past, holding a prominent place in ancient legends, stories, artworks, and in books by various authors. Perhaps the most well-known reference to spirit doubles or 'alter egos' is the doppelgänger, a word still used today to refer to a person that is physically or behaviorally similar to another person.

"Doppelgänger is a German word meaning 'double goer' and refers to a wraith or apparition that casts no shadows and is a replica or double of a living person. They were generally considered omens of bad luck or even signs of impending death—a doppelgänger seen by a person's relative or friend was said to signify that illness or danger would befall that person, while seeing one's own doppelgänger was said to be an omen of death. Some accounts of doppelgängers, sometimes called the 'evil twin,' suggests that they might attempt to provide advice to the person they shadow, but that this advice can be misleading or malicious. They may also attempt to plant sinister ideas in their victim's mind or cause them confusion. For this reason, people were advised to avoid communicating with their own doppelgänger at all costs."

A doppelgänger is a type of wraith who takes on the exact appearance of someone. In German folklore, they serve as an omen of death.

Consider these words, too, from Tom Little at *Atlas Obscura*, who traced the doppelgänger phenomenon from ancient Egypt through the Victorian era. Having mentioned sightings in relation to English writers Percy Bysshe Shelley and Mary Shelley, Little added: "Doppelgänger encounters continued in the U.S., with several cases following a pattern of three sightings preceding death. Soon after his election in 1860, Abraham Lincoln saw his reflection doubled in the mirror, with one face beside the other with a ghostly pallor. He tried to show his wife the apparition, which appeared two more times when she was not present. While Mary Todd was at first worried about this behavior, she took the vision as a sign that he would serve two terms but would die before the end of the second.

"Lincoln is far from the only American to meet their double in the 1800s. The antebellum South was home to numerous accounts of fateful sightings, each under similar circumstances. Linda Derry, site director at the Old Cahawba ghost town in Alabama, is a curator of folklore originating from that region. She has uncovered several cases with similar circumstances as Lincoln's sightings."

And how about this from *Paranormal Guide*? It states: "Although the combining of the words to form the term is relatively recent, a little over two centuries old, the idea of a spiritual, ghostly or demonic double (we'll just use the term 'ghostly' from here) of living people have existed for millennia. These doubles may at times be seen by others as performing a person's actions before the real person makes them, or they may be a shadow, performing the same movements, but after they have happened. They may also be seen in one's reflection; however, the reflection is facing away. Much of the time a doppelgänger is viewed as an omen for a tragedy, illness or death of the person who is copied. If someone sees their own ghostly double it generally bodes very badly for them, and a number of quite famous people have had the ghastly experience."

Now let's dig further into the issue of doppelgängers and of what just might be robots. Janey Tracey, in a 2015 article for the website *Outer Places*, states: "Traveling to the past at the risk of destroying oneself is a common staple of science fiction, not to mention the central concern behind the grandfather paradox. In real life, we still have no idea if time travel is really possible, but a physicist determined in a recent study that if it were possible to travel to the past, *it would likely involve the creation of a pair of ghostly twins that ultimately annihilate each other*" (italics mine).

One of those who suspected that the doppelgänger phenomenon may have been caused by time-traveling doubles was the late paranormal authority and author Brad Steiger. He had a very good reason to suspect it was the answer to a mystery. And why might that be? Well, I'll tell you: Brad had a fascinating story of his own to tell. He shared the story with me just a few years before he died. It goes as follows:

"In the following cases I suspect a human agency involved in a strange campaign that was conducted regarding Steiger imposters who spoke at various conferences around the United States. On occasions the imposters

allegedly conducted themselves very well, thus making the whole enterprise of counterfeit Steigers a seemingly futile project. On other occasions, the imposter's assignment was quite obviously to taint my reputation.

Author and paranormal authority Brad Steiger believed that doppelgängers could be the result of time travel.

"On an unfortunate number of occasions, I received letters complaining of my outrageous and insulting behavior while speaking at a conference. There were claims that I had openly berated my audience, calling them stupid for accepting the very premise of UFOs. A close friend happened to arrive on the scene after one pseudo-Steiger had departed and tried his best to assure the sponsors of the event that the rowdy, disrespectful speaker could not have been the real Brad. In his letter, my friend warned me that he had visited a number of lecture halls where the imposter had damned his audiences. 'Someone seems out to damage your reputation,' he advised.

"In a most bizarre twist, dozens of men and women have approached me at various lectures and seminars, congratulating me about the manner in which I bested Dr. Carl Sagan in debate. The event allegedly occurred after a lecture when I happened to bump into the great scientist in a restaurant. The eatery, according to the witnesses, was crowded with those who had attended the seminar, and they egged on a debate between myself and Dr. Sagan. I mopped up the floor with him, countering his every argument against the reality of UFOs.

"The truth is that I never met Dr. Sagan; therefore, neither had I ever debated him. But from coast to coast, there are those who claim to have witnessed my triumphal bout. Even more individuals claim to have been in the audience when I delivered a rousing message from the Space Brothers in Seattle. Regardless of how often I deny that I was not in Seattle at that time and have never channeled the Space Brothers, those who were at that event are puzzled why I would deny my eloquence."

Multiple Brad Steigers wildly careening their collective way throughout the timelines? Could be.

Roswell, Robots, and UFOs

In July 1997, one of the most controversial UFO-themed books was published. Its title: *The Day after Roswell.* Its author: Philip J. Corso. Its ghostwriter: Bill Birnes of the television show *UFO Hunters*. Simon and Schuster, the publisher, provided the following blurb for the book:

"A breathtaking exposé that reads like a thriller, *The Day after Roswell* is a stunning depiction of just what happened in Roswell, New Mexico all those years ago and how the effects of this mysterious unidentified aircraft crash are still relevant today. Former member of President Eisenhower's National Security Council and the Foreign Technology Desk in the United States Army, Colonel Philip J. Corso was assigned to work at a strange crash site in Roswell in 1947. He had no idea that his work there would change his life and the course of history forever. Only in his fascinating memoir can you discover how he helped removed alien artifacts from the site and used them to help improve much of the technology the Army uses today, such as circuit chips, fiber optics, and more. Laying bare the United States government's shocking role in the Roswell incident—what was found, the cover-up, and more—*The Day after Roswell* is an extraordinary memoir that not only forces us to reconsider the past, but also our role in the universe."

Besides the technological artifacts, Corso claimed that he saw the bodies of the dwarfish, black-eyed things from the Roswell crash and maintained they were not aliens *per se* but rather biological robots created by an alien race that we, the human race, had never yet seen before. What is more, he claimed that same race of robots were time travelers. Some of this data, specifically as it related to the robotic assertion, was similar to that of a man named Nigel Kerner. He was the author of *The Song of the Greys* and *Grey Aliens and the Harvesting of Souls*—the "Greys" being a popular term in UFO research for "aliens."

Corso claimed that he saw the bodies of the dwarfish, black-eyed things from the Roswell crash and maintained they were not aliens per se but rather biological robots created by an alien race that we, the human race, had never yet seen before.

As for the synopsis of *Grey Aliens and the Harvesting of Souls*, it reads as follows: "In 1997 Nigel Kerner first introduced the notion of aliens known as Greys coming to Earth, explaining that Greys are sophisticated biological robots created by an extraterrestrial civilization they have long since outlived. In this new book Kerner reveals that the Greys are seeking to master death by obtaining something humans possess that they do not: souls. Through the manipulation of human DNA, these aliens hope to create their own souls and, thereby, escape the entropic grip of the material universe in favor of the timeless realm of spirit.

"Kerner explains that genetic manipulation by the Greys has occurred since biblical times and has led to numerous negative qualities that plague humanity, such as violence, greed, and maliciousness. Racism, he contends, was developed by the aliens to prevent their genetic experiments from being compromised by breeding with others outside their influence. Examining historical records, Kerner shows that Jesus, who represented an uncorrupted genetic line, warned his disciples about the threat posed by these alien interlopers, while Hitler, a pure product of this alien intelligence, waged genocide in an attempt to rid Earth of all those untouched by this genetic tampering.

"Despite the powerful grip the Greys have on humanity, Kerner says that all hope is not lost. Greys exist wholly in the material world, so if we follow the spiritual laws of reincarnation and karma, aiming for enlightenment and rising above the material—a state the Greys are unable to reach—we can free ourselves from their grasp."

This was expanded on by Kerner's associate, Danielle Silverman, in personal correspondence: "Are these alien visitors a demonic deception created by occultists piercing a hole into the demonic world or are they the actual tangible result of a technology superior to ours? Are they Golems created by esoteric means or are they technologically created entities? I have to say the most probable for me is the technological scenario. As far as I can see even our own technology is not far from creating the kinds of artificial intelligence and bio-technology necessary to make something like the Greys.

"His view is that the dangers lie in the creation of artificial intelligence not in dabbling in the occult. As I'm sure you would have seen yourself, the whole thing could so easily be the reverse of what the Collins Elite claim: i.e., the demons, goblins etc. of the past could have been the Greys of today but reported as metaphysical creations because those who witnessed them had no references against which to describe the technological features of that which they saw. In other words our current technology would be seen as magic by the people of the past.

"If you recall, Nigel Kerner's first book, *The Song of the Greys,* posed the very interesting idea that we had something far more precious about us as natural living beings that would interest a super intelligent alien cartel just a little more than our ability to mine for precious metals.

"Nigel went on to delineate a fascinating concept, explaining that a 'soul' might be a derivational information field that comes out of a natural cadence that came into the Universe with the big bang. This field holds the power to maintain information in what he called a morphogenetic electro-spatial field with an eternal scope of existence in whatever form circumstance allows. The soul is thus an ancestrally contiguous and coherent mechanism for holding information. Nigel argued that if the Greys are an artificially created and manufactured roboidal form then they would find our facility for 'soul' an analogue of their own creators. This would be

One theory about the alien Greys is that they are robots of some kind, but not in a mechanical sense. Instead, they might be synthetic biological forms that are programmable.

an analogue that they would want to try to combine with their own natures themselves.

"The 'Greys,' he argued, were synthetic biological programmed artificially intelligent machine entities sent out as super intelligent robots far more advanced than those we now ourselves send out to explore the Universe. They are however subject to wear and tear as indeed are all atomically derived entities through the second law of thermodynamics which drives them into greater and greater states of decay with time. A non-atomic property, however, such as 'soul,' is immune to the second law. Thus, this property was their ultimate goal, their 'holy grail,' so to speak. It was a property that they could never know or understand in its own terms as purely atomic artificial creations. Their only perception of it was the effect it has on atoms—the difference, in other words, between a naturally living entity and their artificial state.

"Ironically, what these artificially intelligent entities are seeking is eternal survivability. The Greys, and any physical life that they might reconstitute from the DNA codes they carry on behalf of their creators, cannot access the state beyond death. They have no 'soul'; they cannot be born. They are trying to use us as a bridge into 'soul' writing their programs into us through genetic engineering and implantation so that through us they can tap into an eternal existence. Paradoxically, they are trying to achieve the impossible, but as purely physical creations without a 'soul' they cannot know this.

"It is Nigel's suggestion that the fastest highways to travel through the universe lie in what he calls the 'fields of death,' which he identifies as the zero-point field in the space between atoms. Thus, when we die, we can reincarnate at other planetary locations. The vast distances involved in physical space travel and the speed-of-light-beating technology that this involves with enormous G-forces are only survival by artificial roboidal bodies fortified by mercury and other heavy metals. Thus, the Greys are also searching out our ability to transcend and conquer the limits of space and time when we die.

"I hope this is not too long winded, Nick, but after reading your book and the fascinating research you have done, the alternative to fundamentalist Christianity was blazing in my mind and I wanted to share it with you. As you know, Nigel does not feel that any religion corners the market on the truth and points to the wisdom of all the great religious teachers. He does, however, identify the Shroud of Turin and the processes involved in forming this unique and remarkable image as evidence of the power of soul that the Greys are seeking out in its ultimate form. Thus, he suggests it was they who met Jesus in the desert and took him up to see 'all the cities of the world'—a viewpoint only possible from a spaceship!"

There's another aspect to the Corso controversy. *Time* says: "*The Day after Roswell* numbers among its many revelations the claim that ever since 1947, when the Roswell crash put the military on alert, the U.S. government has been fighting 'the real cold war' against what Corso says the military calls EBEs, or extraterrestrial biological entities. Fortunately, it turns out, Ronald Reagan's Strategic Defense Initiative tipped the balance of power."

As Corso writes, "[The U.S. and U.S.S.R.] both knew who the real targets of SDI were.... When we deployed our advanced particle-beam weapon and tested it in orbit for all to see, the EBEs knew, and we knew that they knew that we had our defense of the planet in place."

The U.S. Department of State, in a report online in its archive collection, provides us with a detailed background on SDI/Star Wars:

"On March 23, 1983, in a televised address to the nation, U.S. President Ronald Reagan announced his intention to embark upon groundbreaking research into a national defense system that could make nuclear

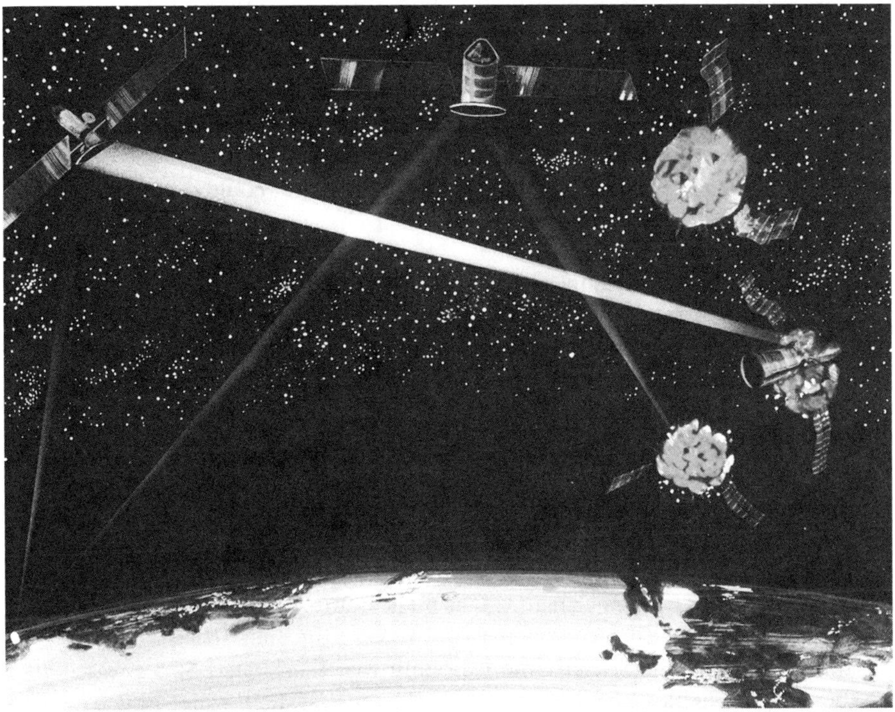

President Ronald Reagan proposed the Strategic Defense Initiative (SDI) as a network of weapons protecting the United States from long-range missiles from a country such as the Soviet Union. However, perhaps the intent was to protect us from a completely alien threat.

weapons obsolete. The research took a number of forms which collectively were called the Strategic Defense Initiative, or SDI.

"The heart of the SDI program was a plan to develop a space-based missile defense program that could protect the country from a large-scale nuclear attack. The proposal involved many layers of technology that would enable the United States to identify and destroy automatically a large number of incoming ballistic missiles as they were launched, as they flew, and as they approached their targets. The idea was dependent on futuristic technology, including space-based laser systems that had not yet been developed, although the idea had been portrayed as real in science fiction. As a result, critics of the proposal nicknamed SDI 'Star Wars' after the movie of the same name.

"There were several reasons why the Reagan Administration was interested in pursuing the technology in the early 1980s. One was to silence domestic critics concerned about the level of defense spending. Reagan described the SDI system as a way to eliminate the threat of nuclear attack; once the system was developed, its existence would benefit everyone. In this way, it could also be portrayed as a peace initiative that warranted the sacrifice of funds from other programs. Privately, Reagan was quite adamant that the goal of U.S. defense research should be to eliminate the need for nuclear weapons, which he thought were fundamentally immoral. In terms of the Cold War conflict with the Soviets, a successful defense system would destroy the Soviet ability to make a first strike, which in turn would undermine the USSR's ability to pose a threat to the United States at all. So, success in this area, supporters of SDI argued, could potentially also bring an end to the Cold War.

"Criticism of the SDI initiative was widespread, however, and it took several forms beyond general skepticism about the feasibility of the technology. First, research and development for such a complicated project inevitably came with a very high price tag. Many critics of SDI wondered why the Reagan Administration was willing to spend so much money on a defense system that might never work and expressed alarm that the funding for SDI came at the cost of social programs like education and health care. Moreover, there was no way to test such a system without exposing the world to a very dangerous attack. Second, the very idea of guarding against nuclear attack struck at the heart of the theory of deterrence. If one nuclear power no longer had to fear nuclear attack, then there would be no fear of retaliation to stop it from making the first strike against another. In fact, if the Soviet Union thought that the United States was on the verge of deploying a comprehensive defense system, some argued, it might feel forced to attack before the United States could complete the system; this possibility meant that developing the system could actually contribute to U.S. insecurity, not the other way around. Third, critics both in the United States and around the world called the SDI initiative a clear violation of the 1972 Antiballistic Missile Treaty. That treaty had committed the United States and the Soviet Union to refrain from developing missile defense systems in order to prevent a new and costly arms race. The Strategic Defense Initiative appeared to be a missile defense system by another name.

President Ronald Reagan is shown here (left) meeting with General Secretary Mikhail Gorbachev of the U.S.S.R. in Geneva, Switzerland in 1985. By this time, the Soviets were having economic problems, forcing them to make some concessions to United States demands, such as the SDI system.

"The Soviet Union expressed its concerns about SDI almost as soon as it learned of it, and the prospect of the United States developing the defense system thus became a hindrance in the pursuit of future arms negotiations between the two powers. Soviet leader Mikhail Gorbachev linked his demands that the United States drop SDI to the negotiations for the Intermediate-range Nuclear Forces Treaty (INF Treaty) and the Strategic Arms Reductions Talks (START). Over the course of the 1980s, Reagan's refusal to give up SDI became the sticking point that prevented the two countries from reaching a deal on other arms control measures, and it was only when the two sides agreed to delink defense and intermediate-range forces discussions that they managed to sign the INF Treaty. START was completed after Reagan left office, and government commitment to the SDI project waned."

Roswell is without doubt the most famous UFO case of all. Second, for sure, is the alleged UFO landing at Rendlesham Forest, Suffolk, U.K., in December 1980.

Roswell is without doubt the most famous UFO case of all. Second, for sure, is the alleged UFO landing at Rendlesham Forest, Suffolk, U.K., in December 1980. To understand exactly what occurred in the woods on those fateful nights, it's important that we go back to the beginning to review (a) the events themselves; and (b) a near-legendary memo on the incidents that was carefully prepared by Lieutenant Colonel Charles Halt. At the time, he was the deputy base commander at RAF Bentwaters. On January 13, 1981, Halt prepared the following report; it was sent to the U.K. Ministry of Defense (MoD) for scrutiny. It gives a fairly brief—but certainly detailed—account of what happened. There's something else, too. Namely, a connection to what Philip Corso claimed: that there was a time-travel aspect to all of this. Before we get to that, let's get back to Halt's memo:

> 1. Early in the morning of 27 Dec 80 (approximately 0300L) two USAF security police patrolmen saw unusual lights outside the back gate at RAF Woodbridge. Thinking an aircraft might have crashed or been forced down, they called for permission to go outside the gate to investigate. The on-duty flight chief responded and allowed three patrolmen to proceed on foot. The individuals reported seeing a strange glowing object in the forest. The object was described as being metallic in appearance and triangular in shape, approximately two to three meters across the base and approximately two meters high. It illuminated the entire forest with a white light. The object itself had a pulsing red light on top and a bank(s) of blue lights underneath. The object was hovering or on legs. As the patrolmen approached the object, it maneuvered through the trees and disappeared. At this time the animals on a nearby farm went into a frenzy. The object was briefly sighted approximately an hour later near the back gate.
>
> 2. The next day, three depressions 1.5 inches deep and 7 inches in diameter were found where the object had been sighted

on the ground. The following night (29 Dec 80) the area was checked for radiation. Beta/gamma readings of 0.1 milliroentgens were recorded with peak readings in the three depressions and near the center of the triangle formed by the depressions. A nearby tree had moderate (0.05–0.07) readings on the side of the tree toward the depressions.

3. Later in the night a red sun-like light was seen through the trees. It moved about and pulsed. At one point it appeared to throw off glowing particles and then broke into five separate white objects and then disappeared. Immediately thereafter, three star-like objects were noticed in the sky, two objects to the north and one to the south, all of which were about 10 degrees off the horizon. The objects moved rapidly in sharp, angular movements and displayed red, green and blue lights. The objects to the north appeared to be elliptical through an 8–12 power lens. They then turned to full circles. The objects to the north remained in the sky for an hour or more. The object to the south was visible for two or three hours and beamed down a stream of light from time to time. Numerous individuals, including the undersigned, witnessed the activities in paragraphs 2 and 3.

Charles I. Halt, Lt Col, USAF

Deputy Base Commander

Two of the most important and credible figures in this entire story are John Burroughs and Jim Penniston. Burroughs was in the U.S. Air Force for more than a quarter of a century, working in law enforcement. As for Penniston, he entered the USAF in 1973. At the time all hell broke loose in Rendlesham Forest, Penniston was a senior security officer. Both men had startling encounters in those December nights. Penniston actually touched the whatever-it-was; something that he now believes caused him to receive a binary code message that was, essentially, downloaded into his mind.

Techopedia explains what, precisely, binary codes are: "Binary code is the most simplistic form of data. It

A photograph of the site at Rendelsham Forest where a triangular-shaped UFO was reportedly spotted in 1980.

is represented entirely by a binary system of digits consisting of a string of consecutive zeros and ones. Binary code is often associated with machine code in that binary sets can be combined to form raw code, which is interpreted by a computer or other piece of hardware."

In a May 6, 2018, article for the *Mysterious Universe* website, titled "Aliens: Us from a Future Time," I wrote in part: "Formerly of the U.S. Air Force, and one of the key military players in the famous UFO encounter at Rendlesham Forest, Suffolk, England in December 1980, Sergeant Jim Penniston—in 1994—underwent hypnotic regression, as part of an attempt to try and recall deeply buried data relative to what occurred to him during one of Britain's closest encounters. Very interestingly, and while under hypnosis, Penniston stated that our presumed aliens are, in reality, visitors from a far-flung future. That future, Penniston added, is very dark, in infinitely deep trouble, polluted and where the human race is overwhelmingly blighted by reproductive problems. The answer to those same, massive problems, Penniston was told by the entities he met in the woods, is that they travel into the distant past—to our present day—to secure sperm, eggs and chromosomes, all as part of an effort to try and ensure the continuation of the severely waning human race of tomorrow."

Could this time-travel aspect referenced by Corso at Roswell have been linked to the reported time-travel angle at Rendlesham Forest, too? If so, then perhaps there is a good chance that Penniston's future humans were themselves robots.

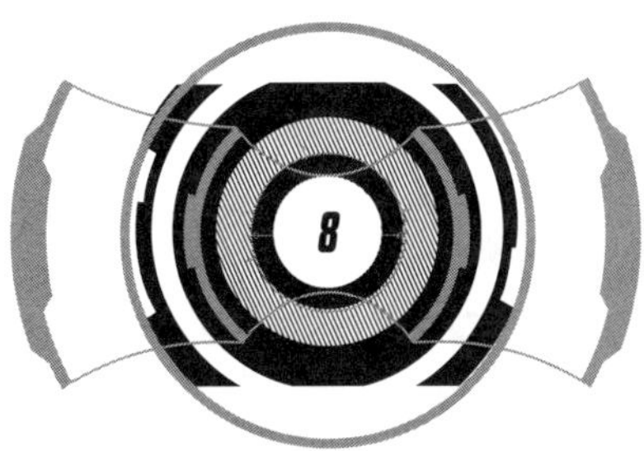

Men in Black: Biological Robots or Alien Overlords?

Make mention of the Men in Black to most people and doing so will likely provoke images of Will Smith and Tommy Lee Jones. After all, the trilogy of *Men in Black* movies were phenomenally successful and brought the subject to a huge, worldwide audience. Outside of ufology, most people assume that the Men in Black were the creations of Hollywood. This, however, is very wide of the mark: in reality, the movies were based upon a short-lived comic-book series that was created by Lowell Cunningham in 1990. Most important of all, the comic books were based on real-life encounters with the MIB—which date back decades.

In fact, in the movies, the characters portrayed by Jones and Smith are known as J and K. There is a good reason for that: they are the initials of the late John Keel, who wrote the acclaimed book *The Mothman Prophecies* and who spent a lot of time pursuing MIB encounters, particularly in the 1960s and 1970s. In that sense, the producers of the *Men in Black* movies and comic books were paying homage to Keel. Now let's get to the heart of the matter: namely, the real Men in Black, not those of Hollywood. Who are they? Where do they come from? What is their agenda? If there is one thing we can say for sure when it comes to the matter of the MIB, it's that they are the ultimate controllers—they threaten, intimidate, and terrify into silence those who they visit. Let's see how the mystery began.

It was in the early 1950s that a man named Albert Bender created a UFO research group called the International Flying Saucer Bureau.

The group was based in Bender's hometown of Bridgeport, Connecticut. Bender quickly became enthused by the UFO phenomenon when it kicked off in earnest in the summer of 1947, with Kenneth Arnold's acclaimed and now-legendary sighting of a squadron of UFOs over the Cascade Mountains. The world was changed, and so was Albert Bender.

Ufologist Albert Bender was the founder of the short-lived International Flying Saucer Bureau and the first to report on the Men in Black.

As a result of the establishment of the IFSB, Bender found himself inundated with letters, phone calls, and inquiries from people wanting information on the UFO enigma. Bender was pleased to oblige, and he created his very own newsletter—*Space Review*. The publication was regularly filled with worldwide accounts of UFO activity, alien encounters, and sightings of flying saucers. On the worldwide issue, it's worth noting that so popular was Bender's group and magazine that he found himself inundated with letters from all around the planet: communications poured in from the U.K., from Australia, from South America, and even a few from Russia. Bender was on a definitive high: the little journal that he typed up from his attic room in the old house in which he lived was suddenly a major part of ufology. It's most curious, then, that in the latter part of 1953, Bender suddenly shut down the International Flying Saucer Bureau, and he ceased the publication of *Space Review*. Many of Bender's followers suspected that something was wrong—very wrong. They were right on the money, as it happens.

When Bender brought his UFO-themed work to a hasty end, a few close friends approached him to find out what was wrong. After all, right up until the time of his decision to quit, he was riding high and had a planet-wide following. It couldn't have gotten much better for Bender. So his decision to walk away from all things saucer-shaped was a puzzle. One of those who wanted answers was Gray Barker. A resident of West Virginia and both a writer and a publisher who also had a deep interest in UFOs, Barker had subscribed to *Space Review* from its very first issue and had developed a good friendship and working relationship with

Bender—which was an even bigger reason for Barker to question Bender's decision.

At first, Bender was reluctant to share with Gray Barker his reasons for backing away from the subject that had enthused him for so long, but he finally opened up. It turns out—Barker wrote in his 1956 book on the Bender affair, *They Knew Too Much about Flying Saucers*—that Bender had been visited by a trio of men, all dressed in black, who warned him not only to keep away from the subject but to completely drop it. As in forever. Somewhat of a nervous character at the best of times, Bender hardly needed telling once. Well, yes, actually, he did: despite having the fear of God put in him, Bender at first figured that what the Men in Black didn't know wouldn't hurt them. So, despite the initial threat, Bender chose to soldier on. It was a big, big mistake. When the MIB realized that Bender had not followed their orders, they turned up the heat to an almost unbearable level. Finally, Bender got the message.

The scenario of a mysterious group of men in black suits terrorizing a rising UFO researcher would make for a great book, thought Barker—which it certainly did.

For Gray Barker—who recognized the dollar value in the story of his friend—this was great news, in a strange way. The scenario of a mysterious group of men in black suits terrorizing a rising UFO researcher would make for a great book, thought Barker—which it certainly did. Hence Barker's 1956 book. The problem was that although Albert Bender somewhat reluctantly allowed Barker to tell his story, Bender didn't tell him the whole story. Bender described the three men dressed in black suits and confirmed the threats, but that was about all he would say. As a result, Barker quite understandably assumed that the Men in Black were from the government. He suspected they were from the FBI, the CIA, or the Air Force. Barker even mused on the possibility that the three men represented all of those agencies. When Barker's book was published, it not only caught the attention of the UFO research community of the day, but it also, for the very first time, brought the Men in Black to the attention of just about everyone involved in the UFO issue. A legend was born—a legend that continues.

While Albert Bender certainly didn't lie to Gray Barker, he most certainly did not share with him the full story. In fact, Bender had barely shared the bones of it. There was a good reason for that: the real story was far, far stranger than Barker could have imagined. Yes, Bender was visited by three men in black, but they were not of the kind that the U.S. government of the day might have been expected to dispatch. Rather, they fell into the domain of the supernatural, the paranormal, and the occult.

According to Bender, late one night, after toiling away on his old typewriter in his attic environment, he suddenly started to feel sick. He was overwhelmed by nausea, dizziness, a sense that he might faint, and, most curious of all, the scent of brimstone—or sulfur—that had permeated the room. The odor is associated with paranormal activity and has been for centuries. Bender lay down on the bed, fearful that he might crash to the floor if he did not. In seconds, something terrifying happened: three shadowy, ghostly, spectral beings started to materialize through the walls of Bender's room—yes, through the walls. They didn't need to knock on the door and wait for it to be opened. The silhouette-like trio then started to change: their shadowy forms became more and more substantial, and they finally took on the appearance of regular men. Apart, that is, from several notable differences: their eyes shone brightly, like a piece of silver reflecting the sun. Their skin was pale and sickly looking, and they were thin to the point of almost being cadaverous. They closely resembled the deadly vampires of old that Bender loved to read about in his spare time.

Using telepathy, rather than the spoken word, the three men warned Bender that now was the time for him to leave the UFO issue alone—leave it and never return.

Using telepathy, rather than the spoken word, the three men warned Bender that now was the time for him to leave the UFO issue alone—leave it and never return. Or else. When Bender began to shake with fear, the Men in Black realized that they had gotten their message across, and they duly departed the way they had arrived—through the walls. For days, Bender was in a state of fear that bordered upon hysteria. Finally, though, he thought: why should I quit ufology? After all, I've done so much work; I'm not going to stop now. So Bender didn't stop; he decided to take on the MIB and stand up to their threats. That was a very big mistake on the part of Bender.

In the days ahead, Bender saw the MIB again. On one occasion, late on a Saturday night, Bender was sitting in his local cinema, watching a new movie, when one of the Men in Black materialized in the corner of the cinema, his blazing eyes focused on the terrified ufologist. Bender didn't hang around; he fled the place. On the way home, though, Bender was plagued by the sounds of footsteps behind him, which seemed to be disembodied, as no one was in sight. In the further days ahead, the MIB returned to that old attic, which yet again caused Bender to fall seriously ill. Finally, after another week of all this terror and mayhem, Bender really was done. His time in ufology was over—for the most part, anyway.

Albert Bender's story, as it was told in the pages of Gray Barker's 1956 book *They Knew Too Much about Flying Saucers*, was substantially correct, in the sense that it told of how Bender was visited, threatened, and ultimately driven to leave ufology. Through no fault of his own, though, Barker was unaware of the supernatural aspects of the story and assumed that Bender had become a victim of the U.S. government. Eventually, though, Bender came clean with Barker. Far from being disappointed, Barker was overjoyed, chiefly because he realized that he could spin the Bender saga into yet another book, which is exactly what happened. This time, though, Barker let Bender write the story himself. Yes, despite being warned away from the flying saucer issue by the Men in Black, Bender somewhat reluctantly reentered the scene and wrote his very own story: *Flying Saucers and the Three Men*, which Gray Barker eagerly published in 1962. Many people in ufology were put off by the overly supernatural aspects of the story, and as a result, the book was relegated to the realm of obscurity for many years.

It's interesting to know, though, that behind the scenes there was another group of men in black suits—and black fedoras—who were secretly following the Bender saga. It was none other than the FBI. In other words, although the FBI were not literally Bender's MIB, the FBI certainly wanted to find out who they were. Thus, in a strange way, there were now two groups of MIB, both distinctly different: the supernatural ones encountered by Bender and the MIB of government officialdom. The provisions of the Freedom of Information Act have shown that both Albert Bender and Gray Barker had files opened on them. Those same files make it clear that none other than the legendary FBI boss J. Edgar

Hoover had ordered one of his special agents to get hold of a copy of Gray Barker's *They Knew Too Much about Flying Saucers.*

Bender spent his later years managing a motel in California. He passed away at the ripe old age of 94.

After promoting his book, Bender yet again walked away from the UFO issue. This time, it was for good. Bender died in March 2016, at the age of 94, in California.

In the years that followed Bender's encounters, the U.S. government would become determined to uncover the truth of the MIB. During the course of his research into the issue of the Men in Black, John Keel arranged a meeting with one Colonel George P. Freeman of the U.S. Air Force. Keel's interest was driven by the fact that Colonel Freeman had circulated a memo throughout the Air Force ordering everyone to be on guard for the Men in Black. Colonel Freeman's memo read as follows:

"Mysterious men dressed in Air Force uniforms or bearing impressive credentials from government agencies have been silencing UFO witnesses. We have checked a number of these cases, and these men are not connected to the Air Force in any way. We haven't been able to find out anything about these men. By posing as Air Force officers and government agents, they are committing a federal offense. We would sure like to catch one—unfortunately the trail is always too cold by the time we hear about these cases, but we are still trying."

Only a few weeks after Colonel Freeman's memo was widely circulated, there was this one from Lieutenant General Hewitt T. Wheless, also of the U.S. Air Force: "Information, not verifiable, has reached Hq USAF that persons claiming to represent the Air Force or other Defense establishments have contacted citizens who have sighted unidentified flying objects. In one reported case, an individual in civilian clothes, who represented himself as a member of NORAD, demanded and received photos belonging to a private citizen. In another, a person in an Air Force uniform approached local police and other citizens who had sighted a UFO, assembled them in a school room and told them that they did not see what they thought they saw and that they should not talk to anyone about the sighting. All military and civilian personnel and particularly information officers and UFO investigating officers who hear of such reports should immediately notify their local OSI offices."

It was this period of interest in the MIB on the part of the government that led to an extraordinary, and almost surreal, development.

Although the U.S. government had no real idea of who or what the real Men in Black were, there was a realization on the part of the government that the phenomenon of the MIB could be used to the advantage of the likes of the NSA, the CIA, and military intelligence. It wasn't just the MIB who wanted UFO witnesses silenced; the government did, too, but the government was concerned about threatening UFO witnesses—American citizens, in other words—and being outed in the process.

So the government came up with an ingenious idea: it created a group within the heart of officialdom whose job it would be to keep people away from the really important parts of the UFO phenomenon. Threats, silencing, and intimidation were the orders of the day, but how was that successfully achieved? By having their secret agents dress and act like the real MIB who had terrorized Albert Bender and had intruded on the life of Brad Steiger. In other words, they wore black suits, black sunglasses, and black fedoras, and they acted in a distinctly odd, emotionless fashion. The government really did not know (and probably still does not know) who or what the MIB really were, but that same government knew that it could exploit the phenomenon to its distinct advantage. Dressing as the MIB could offer government agents an ingenious form

The government really did not know (and probably still does not know) who or what the MIB really were, but that same government knew that it could exploit the phenomenon to its distinct advantage.

of camouflage. And so, it did. It was a case of using fear to provoke the ultimate form of control.

As the 1960s became the 1970s, and then the 1980s and the 1990s, and now the twenty-first century, the issue of the existence of two different types of MIB—government agents and something supernatural—continued.

Let's now take a look at some latter-day cases. In 2011, the following extraordinary account was provided to me by a British man named Tim Cowell, a freelance videographer who has been filming professionally since 2008. He has a bachelor of arts degree in film, television, and advertising from the University of Wales, Aberystwyth. His filming credits include the FashionTV network, corporate businesses, the Wrexham council, the education sector, and various documentaries. Alongside his freelance work, he is pursuing his second degree in creative media technology at Glyndwr University. He also volunteers his photography skills to the Wrexham County Borough Museum and Archives.

Cowell's account demonstrates that whoever, or whatever, the Men in Black may be, they were as active in the 1990s as they were when the likes of poor Albert Bender was being terrorized back in the early 1950s. Notably, as our correspondence progressed, Tim revealed that—MIB aside—he had lifelong experience of strange phenomena, including encounters of both a ghostly and a ufological kind. Just like Albert Bender, in fact….

"Dear Mr. Redfern,

"The reason I am writing to you is with regards to a strange experience I had back in 1997 when I was 17 years old. Whether you may be able to shed light on my experience I'm not sure, but I came across your name and 'real men in black' article on the web a few moments ago and felt that your expertise on the subject might lift a nagging uncertainty that I have had for fourteen years.

"Firstly I would like to say that I have not read your 'Men in Black' book as yet (I do intend to) but I do have an interest in the unexplained and have read many books on these subjects since I have had multiple strange experiences in my past and present. That being said, the experience I wish to convey to you has not been contaminated with any theories of others or my own.

"I am very open-minded but at the same time possess a healthy skepticism with any unexplained phenomena. However, I have not found any logical reason for what I am about to tell you (although there is always the possibility that there is one). The following account is complete truth and

I have not embellished any part of it. All I hope is that you might have an explanation for what happened, be it strange or mundane, as I am uncertain as to whether this account depicts the behavior of the 'Men in Black.' At the time of the experience I was 17 years of age and was 'bunking off' [skipping class] from a college lecture to meet my then girlfriend later that afternoon....

"My Account: I was walking from my college and into town to get a coffee to while away the hour until I caught the bus to my girlfriend. I was young and newly 'in love' and walking quite happily down the main street when I had a strange feeling that I was being followed. This feeling led to an instinct of looking behind me, and as I did, a few feet away, I saw a couple of men close behind. As I looked, they both emitted a 'blank' smile. Being young and—dare I say it—possibly naive, I had the thought that maybe they were from the college and following me because I bunked off. (On reading this whole account you will see that what happened is not the normal procedure any college would take.)

On the surface, the Men in Black do look human, but there is something "off about them" in their cold smiles and blank stares.

"After I witnessed this 'blank smile' I continued to walk at my normal pace down the long main street towards my destination. I was now wondering to myself if they smiled at me because I looked at them (the old 'you look at me, so I look at you' scenario). I looked behind me a second time and again they offered, in unison, that same 'blank smile.' I also noticed their appearance and whilst they were not wearing black suits and black fedoras, they were wearing an attire that didn't seem to fit in. Dark brown tweed suits with matching long overcoats and fedora-like hats. Without sounding clichéd (as I now know the usual nonconformity of these guys) they did appear to be from an earlier era than the '90s, to say the least.

"I decided to quicken my pace and noticed that their pace also quickened. Feeling a little paranoid, I quickened my pace again; and again, they also matched my speed. So now I'm almost speed walking towards the cafe to get my coffee. A third look behind me before I entered (the then) 'John Menzies' [a British store-chain] confirmed that they were still walking my way, so I entered the store but waited inside a little for them to pass by. They didn't, so I ventured into the street again, but they had disappeared. I immediately assumed that they had turned off or entered another shop and put it down to myself as being a paranoid college bunker.

"I re-entered the store and proceeded to walk upstairs to the cafe area. It is worth mentioning here that whilst I chose this cafe for its quietness it did always bug me that the cafe attendants rarely gave you enough time to choose what you wanted without being quickly pestered into hurrying up with your order. (The reason for this note will become apparent soon.) Having been quickly served at the counter, I found a place to sit at a table facing the cafe entrance and began to read a letter that my girlfriend had sent me (sickening I know).

"Anyway, a few moments later I looked up from the letter whilst taking a sip of my coffee and froze on the spot. The two distinguished gentlemen were a few feet away at the food counter staring at me blankly. After what seemed like an age of staring one of the men placed a large leather-like satchel that I had not noticed before on the floor. With the other man still looking at me, the other bent down, opened the satchel and pulled out a very large and old-looking camera, complete with large round flash. He proceeded to point the camera directly at me and took my picture. On doing so he placed the camera back in the satchel, and both men turned and slowly walked away towards the stairs.

"Completely in shock and bemused as to what just happened, I was still frozen in place trying to wonder what the hell had just happened. I quickly decided to follow them (the time taken for this decision, taking into consideration the casual speed at which they exited, I calculated that they would still be going down the stairs or at least at the bottom by the time I got to them) and literally ran down the stairs. There was no sign of them, so I decided to go to the store exit first and looked outside, but they weren't anywhere to be seen. I then turned to look into the store again due to the fact that I might have missed them inside and that they would have had to pass me to leave. But again, they were nowhere to be seen.

"One thing that was apparent to me was that whilst they were upstairs by the counter, they were never attended to by the very needy cafe staff—and believe me, they used to pester you. To be honest, without sounding stupid, it seemed like no-one could see them. I know how that sounds but all I can do is explain the account in the same way I experienced it.

"Now, as a 17-year-old bunking off college I was hesitant to tell my mother of this experience (not because of the ludicrous way it would have sounded—she actually took that part in her stride as she has also experienced strange phenomena in her life), but because I thought I would have been grounded for bunking off. Least to say, when I did arrive home later that day I told her the exact same thing I told you now, including why I was not in college.

"The intrigue of my experience swayed the 'grounding' and to this day I have no logical reason as why something like that would happen to me. Obviously with my interest in all things weird becoming increasingly larger over time with other experiences and the ease as to which information about ourselves can be found out via the internet, this aspect couldn't have been the reason for this strange occurrence as I was rarely on the new 'internet' back then.

"Anyway, what happened that day is a mystery and there could be a mundane reason for it. But there are little things that bug me. Why did I feel I was being followed only to see that I was? Why did they seem out of place in both their clothes and their blank demeanor? Why take a picture of me at all, let alone with the most old-fashioned of cameras? And how did they disappear so quickly? Is this the type of behavior that you would deem to be of 'Men in Black' origin?

"I know that account sounded a little 'wacko' but I assure you I am of sound mind. I simply have an experience that I have no answer for. Thank you for taking the time to read this long-winded email and I hope to hear from you soon.

"Kind regards,

"Tim Cowell."

I wrote back to Cowell and asked a few questions regarding the specific location, and received the following in response:

"Hi Nick,

"Thanks for your reply. That experience was in my hometown of Wrexham, North Wales. Like I said,

Wrexham is a charming town in the northeast of Wales, and it is where Tim Cowell says he saw the Men in Black.

it's something that I recall from time to time with a nagging uncertainty as to what it actually was and why.

"Because I have had many paranormal experiences I had wondered if there was any link between them. Most of these have been placed in the more ghostly category but there was an incident when I was even younger that myself and grandma witnessed a UFO sighting. The same night of the sighting I was sharing a bed with my cousin (we were being babysat during a weekend) and when I awoke in the morning my Gran found us 'artificially' laying in the bed . . . myself lying on my back with arms crossed neatly over my chest and my cousin upside down, feet on the pillow and head under the quilt at the bottom. Not a normal way to sleep, and the bed sheets were as if they had been made whilst we were already in them. Strange.

"All through my life I've seen, felt and heard 'ghosts' or whatever in my family home and even more recently encountered paranormal resistance whilst living and working in Malta which required the help of a Catholic priest! I'd love to write a book about my experiences but don't know the first thing about publishing :)

"Anyway, whilst I have and continue to experience strange things I simply had no explanation at all as to who those strange men who followed me were. The only reason I have regained interest in that strange day was thanks to a movie that I had recently watched called *The Adjustment Bureau.* In the same way that smelling a scent can transport you back to a memory, I had the same jolt of surprise when I saw these 'adjustment men' in that movie as their appearance instantly reminded me of that day back in 1997. Thus, thrusting me back onto the internet to try and find anyone with an answer or similar experience to mine.

"And that's when I came across your book, *The Real Men in Black*. I have to say that I do own your book *Cosmic Crashes* and because I enjoyed it and realized that you were the same author, I ventured to ask you your opinion on the matter. Again, thanks for your reply, and I feel better knowing that an author of your caliber and experience on the subject appreciates the weird and wonderful.

"Kind regards,

"Tim Cowell."

Now that we have seen the origins of the Men in Black, let's focus our attention on the possibility of the Men in Black being biological robots. In his excellent catalogue *MIB Encounters,* Gareth Medway includes a story that suggests the Men in Black might be nothing less than biological robots.

Before we get to the matter of the MIB being something far removed from the human race, let's take a look at biological robots in today's world of science. On January 13, 2020, the U.K.'s *Guardian* newspaper ran an article titled "Scientists Use Stem Cells from Frogs to Build First Living Robots." In part, the article stated: "Researchers in the US have created the first living machines by assembling cells from African clawed frogs into tiny robots that move around under their own steam. One of the most successful creations has two stumpy legs that propel it along on its chest. Another has a hole in the middle that researchers turned into a pouch so it could shimmy around with miniature payloads. 'These are entirely new lifeforms. They have never before existed on Earth,' said Michael Levin, the director of the Allen Discovery Center at Tufts University in Medford, Massachusetts. 'They are living, programmable organisms.'"

Do the Men in Black fall into the category of those "programmable organisms"? Let's have a look at what we know about the MIB. We'll begin with that aforementioned story that Gareth Medway highlights. It reads as follows: "The Christiansen family of Wildwood, New Jersey, who had seen a UFO on 22 November 1966, were interviewed by 'the strangest looking man I've ever seen,' wearing a thin black coat, who introduced himself as 'Tiny' from the 'Missing Heirs Bureau.' He spoke in a high, 'tinny' voice, in clipped words and phrases like a computer, 'as if he were reciting everything from memory.' His black trousers were too short, and 'they could see a long thick green wire attached to the inside of his leg, it came up out of his socks and disappeared under his trousers.' John Keel commented that he had not heard of this feature in other MIB cases: 'Was Tiny wearing electric socks? Or was he a wired android operated by remote control?' He departed in a black 1963 Cadillac."

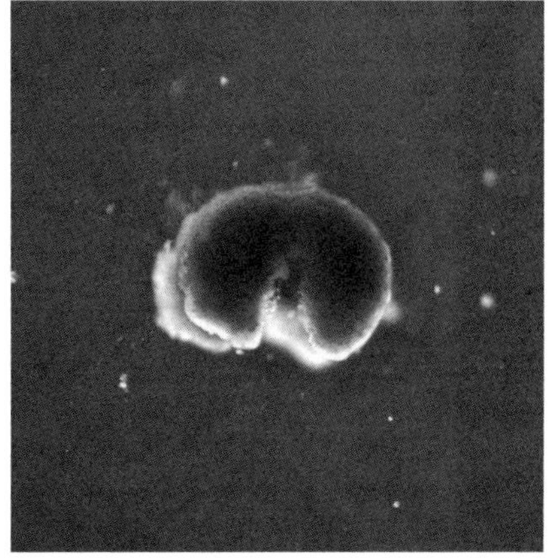

Scientists have managed to create tiny, biological robots out of organic material such as frog skin cells. Called "xenobots," they are synthetic lifeforms that can perform basic functions. Could the Men in Black be more advanced versions of this technology?

Now, further to the theme of this chapter. First and foremost, the MIB are nothing like Will Smith and Tommy Lee Jones. The MIB-themed movies are good fun, but by presenting the Men in Black as the employees of a secret organization that wipes out dangerous aliens, the implication is they are human. They are not. There are so many weird issues surrounding the real MIB phenomena that they strongly suggest the Men in Black are something else entirely. Over the years I have addressed more than a few different theories for what the MIB might be. The list includes time travelers, tulpas/thought-forms, demons, extraterrestrials, interdimensional creatures, and more, but let's stay focused on the biological robot angle.

The fact is that the MIB *do* seem to act in a strange and robotic fashion. There is the matter of their skin. More than a few people who have been confronted by the Men in Black close-up have noticed something very strange. Eerie, in fact. In many cases, their faces appear to be plastic-like—not unlike a nightmarish, hideous mannequin, or a creepy old doll come to life. Moving on, Gareth Medway notes that the MIB talk in a very odd way. He mentioned they seem to recite "everything from memory." I have several such cases in my files. In one case, the witness said the MIB appeared to have no actual understanding of what he/it was saying. The wording, the witness suspected, was programmed. On top of that, there's the matter of the somewhat clumsy, jerky fashion of walking that the MIB have. Certainly, this is not described in every case, but there are more than a few such reports. Finally, there is the angle of food—or, rather, the lack of it. That's right: the MIB have a distinct aversion to food. They don't even seem to know what food is. Perhaps, if the MIB are biological robots, they get their sustenance in a distinctly different way to us. None of this proves that the MIB are biological robots, but specifically as an extremely controversial theory, it's not a bad one.

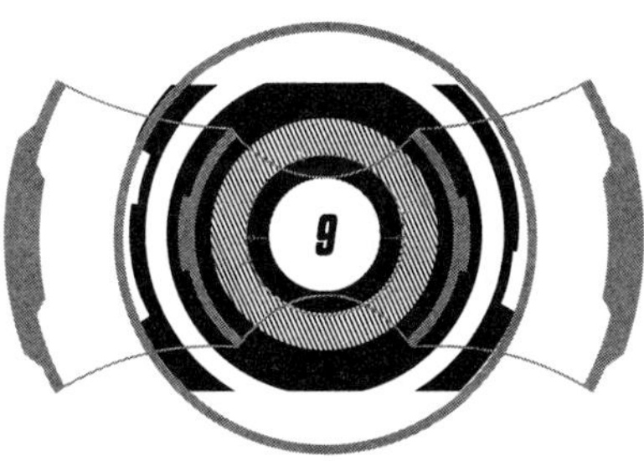

Robotic Spies in the Home

Imagine if you could have a highly advanced piece of technology in your very own home—technology that will answer questions, give you weather updates, set your morning alarm, interact with you, and even listen to you—every single minute of every single day of every single year. Certainly, matters relative to the likes of the weather and your alarm are no big deal, but a device that eavesdrops on every word spoken in the family home? It sounds like the worst nightmare possible. And yet countless numbers of people have already embraced this creepy technology, unaware of the potential violations of privacy that it poses—or, worse still, not even caring about the ways in which their private lives are being opened up to, for example, the intelligence community. Welcome to the world of Amazon Echo and Alexa.

It was almost a decade ago when Amazon first began thinking about creating something along the lines of smart technology that could interact

Helpful robotic assistant or nefarious spy? The Alexa from Amazon can learn a lot about you from the questions and requests you provide to it.

with people and how it might benefit the public. Much of the research and development was undertaken in the heart of Silicon Valley, California.

Of course, as is so often the case with such technology, there is a significant possibility of it being ruthlessly manipulated by those who wish to learn who we are speaking to, what we are doing, and even the content of the conversations that go on in our living rooms, kitchens, bedrooms—in fact, just about everywhere. It sounds like something straight out of a paranoia-filled novel. It's not: it's all too real, and it's a phenomenon that is growing by the minute.

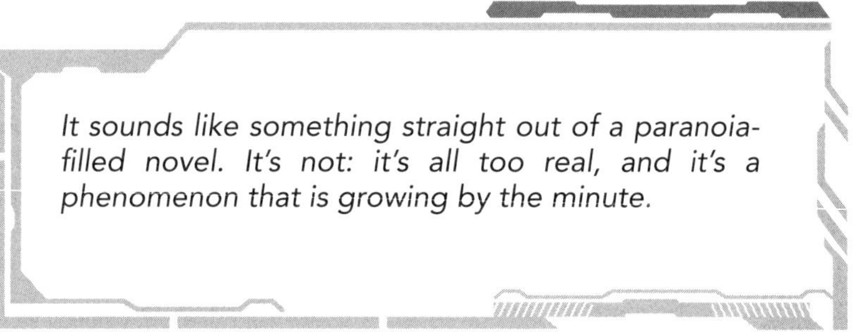

It sounds like something straight out of a paranoia-filled novel. It's not: it's all too real, and it's a phenomenon that is growing by the minute.

In simple terms, Amazon Echo is a device—something called a "smart speaker" that is hooked up to the internet. When you are within range of the device, you can ask it questions—"Who were the Beatles?" "What happened at Pearl Harbor in 1941?" "Who shot JFK?"—and, in quick-time fashion, you will have your questions answered. But who or what is providing the answers? That's where things get kind of creepy. Say "hello" to Alexa, an internet-based personal assistant that voices the answers to your questions and can multitask to degrees that would be impossible for a human being. That's right: Alexa is not a person at the other end of the speaker. Alexa is smart technology taken to the—so far—ultimate degree. Alexa will respond to her (its) name and also to such words as "Computer," "Echo," and "Amazon." Most people, however, prefer to go with Alexa—which gives the ultimate multitasker a degree of personality and gives the user a feeling of interacting with something that is self-aware—which it may well soon become. If it hasn't already, to a degree.

It was in 2015 that Alexa was unleashed upon the public, specifically in the United States, on June 23. Both Canada and the United Kingdom joined Alexa's little club a year later. In theory, there is nothing at all wrong with you having Alexa answer those questions you need answering, but Alexa's role doesn't end there. And this is where things get as complicated as they do controversial. Let's say, for example, that you direct Alexa to play the new song from your favorite band. Another family member does likewise with their favorite band. So does a third. And so on. When Alexa knows which particular music you each individually like, "she" can determine who is in the house just by listening in to what music is playing in the background. In other words, if your teenage daughter likes Taylor Swift, Alexa will understand that it's your daughter in the house and not you. And all because your family has handed over all of its musical tastes to a smart device that is so smart it can figure out who is home and who isn't.

It gets stranger. In 2018, the website *Mysterious Universe* stated the following: "In the latest case of weird Amazon Alexa stories, it's been reported that friendly robot holds some views that aren't, shall we say, accepted by the mainstream. Previously, the Amazon Echo Dot Alexa made headlines when it was creepily laughing at some of its owners. Now it seems like Alexa may be the world's first AI conspiracy theorist. When asked 'Alexa, what are chemtrails?' Amazon's 'intelligent personal assistant' responded by informing the unwitting user that chemtrails are nefarious chemical or biological agents sprayed into the atmosphere by the government. It seems somehow doubtful that Amazon programmed that little tidbit of information into their flagship smart-home intelligence."

So, where did Alexa get her information on chemtrails? *Mysterious Universe* states: "Since this story broke, Amazon's been quiet on how Alexa came up with this answer, saying it was a bug, and quickly announcing they had fixed it. Now, the Amazon Echo Dot Alexa gives the definition of 'contrail' when asked what chemtrails are." What we have here is a case of Alexa using machine-learning algorithms—learning, in essence—and providing answers to questions that should not be in the Echo database. It gets more chilling: in January 2018, Amazon's vice president, Marc Whitten, spoke at the Consumer Electronics Show. He said: "Rolling in things like Alexa, one of the things that we've been learning is that it's not even just necessarily about the facts. One of the big things we're doing

When asked, "What are chemtrails?" Alexa gave a fascinating reply, stating that chemtrails are nefarious chemicals sprayed into the atmosphere by the government.

with Alexa is making sure that she has opinions. What does Alexa think is something that's a good thing to watch?"

Letting Alexa decide what shows we watch? What movies? And Amazon thinks this is a good idea?

Now let's take a look at how Alexa almost became a witness to a murder—albeit in a very strange, alternative, and almost unbelievable fashion.

In February 2016, a man named James Bates of Bentonville, Arkansas, was charged with the murder of one Victor Collins, found dead in his hot tub. There's no doubt that Collins drowned. The big question was: had Collins died accidentally, or was it a case of cold-blooded murder? Bates said that he woke up to find Collins dead. Investigators, though, suspected that Collins had been strangled and drowned. The case was taken to a whole new level when it was realized that Bates owned an Amazon Echo. Could Alexa spill the beans? After all, the one thing that Alexa does better than anything else is to monitor and even record the conversations of the owners and the users. It didn't take the local police long to approach Amazon—with a warrant, no less—and request access to Bates's Echo. This was new and uncharted territory and quickly captured the attention of the media.

Amazon agreed to provide the police with a "record of transactions" but refused to give them any relevant "audio data." Amazon stated: "Given the important First Amendment and privacy implications at stake, the warrant should be quashed unless the Court finds that the State has met its heightened burden for compelled production of such material."

Ultimately, charges against Bates were dropped. The affair was highly instructive and revealing, though, in terms of how, in the future, devices like Echo may well play a role—perhaps even a key and integral role—in what goes on behind closed doors.

In 2018, *The Verge* brought up an important issue related to what extent the National Security Agency might be able to access the likes of Amazon Echo. *The Verge*'s Russell Brandom said: "The NSA has always had broad access to US phone infrastructure, something driven home by the early Snowden documents, but the last few years have seen an explosion of voice assistants like the Amazon Echo and Google Home, each of which floods more voice audio into the cloud where it could be vulnerable to NSA interception. And if so, are Google and Amazon doing enough to protect users?"

It's a question that, as technology advances even further, will be at the forefront of matters relative to the right to privacy versus what government agencies believe they have the right to do in the name of national security.

It's important to note that Alexa is not alone. There's also Siri. Britta O'Boyle at *Pocket-lint* explains: "Siri is a built-in, voice-controlled personal assistant available for Apple users. The idea is that you talk to her as you would a friend and she aims to help you get things done, whether that be making a dinner reservation or sending a message.

"Siri is designed to offer you a seamless way of interacting with your iPhone, iPad, iPod Touch, Apple Watch, HomePod or Mac by you speaking to her and her speaking back to you to find or do what you need. You can ask her questions, tell her to show you something or issue her with commands for her to execute on your behalf, hands-free.

Siri is a digital assistant available on phones that interacts with users in a very human way.

"Siri has access to every other built-in application on your Apple device—Mail, Contacts, Messages, Maps, Safari and so on—and will call upon those apps to present data or search through their databases whenever she needs to. Ultimately, Siri does all the legwork for you."

That sounds great, right? Well, sometimes, yes, but not always. There's an even darker side to the world of what are now known as digital assistants, as Rozita Dara at *The Conversation* explains: "Digital assistants can record our conversations, images and many other pieces of sensitive personal information, including location via our smartphones. They use our data for machine learning to improve themselves over time. Their software is developed and maintained by companies that are constantly thinking of new ways to collect and use our data.

"Similar to other computer programs, the fundamental issue with these digital assistants is that they are vulnerable to technical and process failures. Digital assistants can also be hacked remotely, resulting in breaches of users' privacy.

"For example, an Oregon couple had to unplug their Alexa device, Amazon's virtual assistant, as their private conversation was recorded and sent to one of their friends on their contact list.

"In another incident, a German man accidentally received access to 1,700 Alexa audio files belonging to a complete stranger. The files revealed the person's name, habits, jobs and other sensitive information."

A grave new world? Possibly.

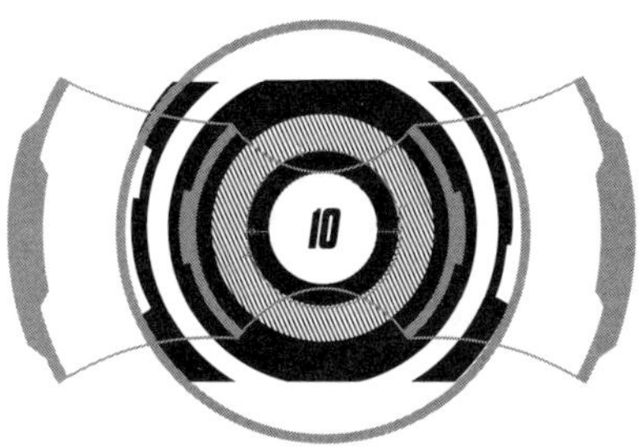

Robocop Comes to Life

Back in July 1987, Orion Pictures produced a movie that quickly became a mega-hit just about everywhere. Starring Peter Weller, Nancy Allen, Miguel Ferrer, and Kurtwood Smith, it was *Robocop*. Situated in Detroit, Michigan, in what, back then, was the future, the movie eerily and graphically managed to predict what, one day, was to come. The story of the blockbuster movie—which reaped in more than $50 million—revolves around a gigantic, powerful organization called Omni Consumer Products. Its staff has the incredible challenge to create a next-generation police officer to deal with the rising chaos and anarchy that has overwhelmed Detroit. If the regular police are no longer able to keep control, then clearly they are in need of something else to work alongside them—or maybe even become the dominating form of law enforcement from then on. As the title of the movie suggests, Peter Weller's character—Officer Alex Murphy, who is transformed into something beyond human—becomes something akin to *The Six Million Dollar Man* meets *The Terminator*. Welcome to the new breed of cop: one that is half-human, half-machine, and takes on the bad guys in fine, fierce fashion.

A witty, exciting, ultraviolent movie, *Robocop*, back in 1987, was just a fine piece of well-made science-fiction. More than three decades later, however, what was seen just as fantasy is on the cusp of becoming all too real. That's right: in the near future, the police may not be entirely human. They may in fact be nothing less than fully functioning cyborgs. If you think I'm exaggerating, I assure you I'm not. Read on.

In 2017, *Forbes* reported: "The Dubai police force is hoping to replace a quarter of its human cops with droid cops by 2030. Should visions of a trigger-happy robot cop be terrorizing your imagination, rest assured this is more of a policeman-shaped help vestibule on wheels than a killer robot with a badge. The uniformed bot cop may be able to read human faces and identify possible miscreants, but it disturbingly lacks a mouth, and its face is a white spread of emptiness embedded with two darkly sunken eyes. Talk about a chilling effect on crime."

EDI Weekly expanded on this incredible development: "Apart from being able to communicate with people, the Robocop can also shake your hand and offer a military salute. It cannot only identify hand gestures from a distance of up to 1.5 meters but can also detect people's emotions and facial expressions. It can figure out when you are happy, sad, smiling or frowning and will itself respond accordingly." They add: "Brig. Khalid Al Razooqi, Dubai Police Director General of smart services officially launched the first Robocop on Sunday, 21st of May, 2017. He said in an interview with 'Gulf News Report' that the Robocop was created to assist and help people in the malls or on the street, the Robocop is the latest smart addition to the force and has been designed to help us fight crime, keep the city safe and improve happiness level."

A cosplayer poses as Robocop at a 2017 event in Indonesia, showing how popular the movie character is even decades after the 1987 film was released. Might modern tech make robocops a reality?

And there are these words from David Moye in *HuffPost*: "Alan Winfield, Professor of Robot Ethics at the University of the West of England, argues there are some serious moral concerns. 'There are big ethical problems,' he told CNN. 'If you're asking a robot to apprehend criminals, how can you be sure that the robot would not injure people?' He adds that guaranteeing a robot can safely intervene in crime-related scenarios 100 percent of the time is 'extremely difficult.' 'Of course, when humans make mistakes they are held to account,' Winfield said. 'The problem is that you can't make a machine responsible for its mistakes.... How do you punish it? How do you sanction it? You can't.'"

We're not quite done. Let's not forget about police dogs. In the near-future they may be robo-dogs. The American Civil Liberties Union (ACLU) addressed this issue in 2012: "The New York Police Department is receiving a lot of attention for testing robot 'dogs,' which it has deployed in several situations, including to deliver food in a hostage situation and to scout out a location where the police feared a dangerous gunman might be lurking. The state police in Massachusetts have also experimented with these robots, as our ACLU colleagues there uncovered, and police in Hawaii have acquired one. What are we to think of these robots from a civil liberties perspective?

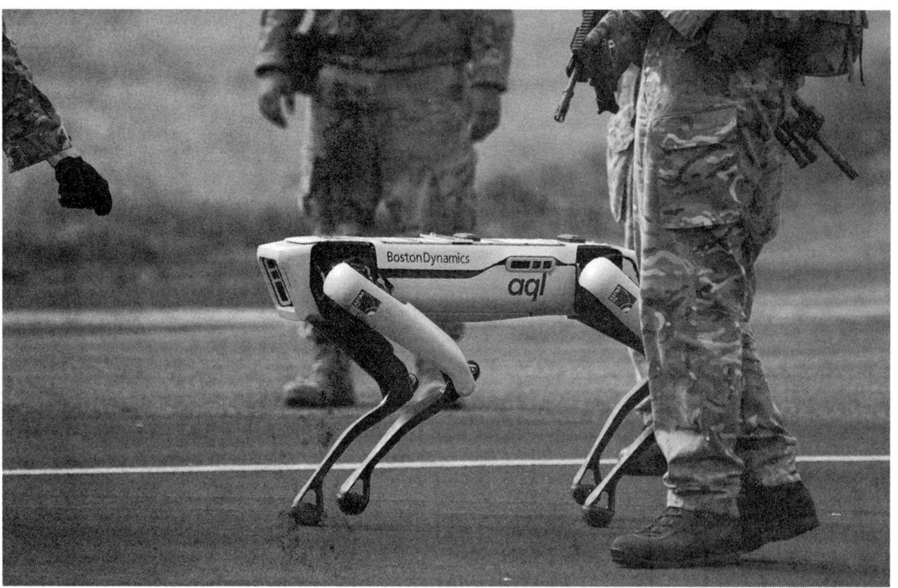

Spot, one of the more recent iterations of doglike robots by Boston Dynamics, debuted in 2016. Much less cumbersome than earlier models, Spot weighs a mere 55 pounds (25 kilograms) and is customizable to perform many tasks. The robot became commercially available in 2020.

"There's definitely something spooky about all the robots made by Boston Robotics [Dynamics], a company that has become famous for videos of its increasingly agile humanoid and animal-oid robots. Add to that spookiness an awareness of our nation's levels of police violence, racism, brutality, and unnecessary killings—and the shameful state of the law that enables those abuses—and this robot police dog, dubbed 'Digidog,' definitely evokes some primal science-fiction fears."

And, finally, muse upon the chilling words of the *Guardian*: "While Boston Dynamics, the creators of the robot dog, have insisted that Digidog will never be used as a weapon, it is highly unlikely that that will remain true. MSCHF, a political art collective, has already shown how easy it is to weaponize the dog. In February they mounted a paintball gun on its back and used it to fire upon a series of art pieces in a gallery. The future of weaponized robot policing has already been paved by the Dallas police department. In 2016, the DPD used a robot armed with a bomb to kill Micah Johnson, an army reservist who served in Afghanistan, after he killed five police officers in what he said was retaliation for the deaths of Black people at the hands of law enforcement. While it was clear that he posed a threat to police, it is very fitting that a Black man would be the first person to be killed by an armed robot in the United States—roughly a year after the white mass shooter Dylann Roof was met with a free burger and police protection."

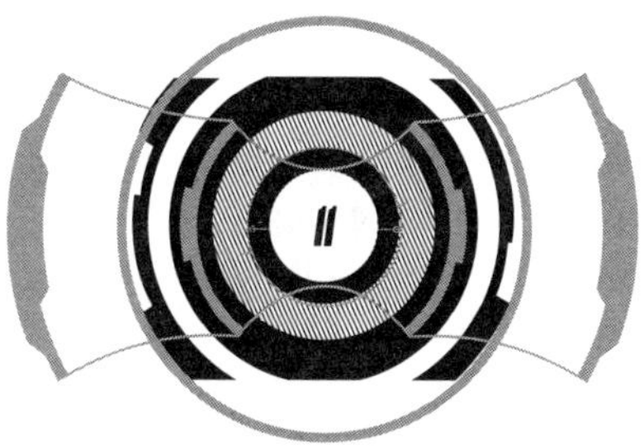

When Your TV Becomes a Technology-Filled Danger

Imagine the scene: It's a Saturday night, and you have a few friends over for drinks and dinner. At one point during the evening, you turn the conversation to the subjects of global politics and the ever-increasing intrusions of the surveillance state. Suddenly, and in unison, everyone in the room goes quiet. It's one of those classic awkward silences. But it's not because your friends have nothing to say on the matter. Quite the opposite, in fact. They actually have a great deal to say. Or, rather, they have a great deal they *wish* they could say. The reason for the silence is as simple as it is sinister and chilling.

Everyone in the room is fearful that your television is listening to their every word. Not only is the TV listening to the words, it's also recording them. Worse still, due to a lack of encryption on the part of the manufacturer of the TV, the National Security Agency is able to access the set and listen in to every single word being said. So, rather than risk the wrath of the NSA and be forever labeled "a person of interest" by the Department of Homeland Security, your friends choose to change the subject and avoid a discussion of matters of a political nature that might one day come back to hurt them—possibly seriously. In stark and almost surreal terms, you are now under the complete control of your television. Even George Orwell didn't see that one coming. Whereas Orwell's classic 1949 book, *1984,* was a work of highly thought-provoking fiction, the issue of entire swathes of the population being controlled and swayed by their primary form of entertainment—the TV—is not fiction. It's bone-chilling fact.

It was in early 2015 that the controversy surrounding this very disturbing issue surfaced to a significant degree. And what an incredible story it proved to be.

Writer Clark Howard stated: "You've heard of the government spying on you and even businesses spying on you. But have you heard of your TV spying on you?! If you're not familiar with 'smart TVs,' they are modern flat-screen TVs with built-in apps allowing you to access online content like Netflix, Hulu Plus, or Amazon Prime much more easily than you would access traditional broadcast content."

As Howard also noted, the one company more than any other that was getting a great deal of justified flak was Samsung. Howard revealed to his readers a specific and important sentence that appears in Samsung's

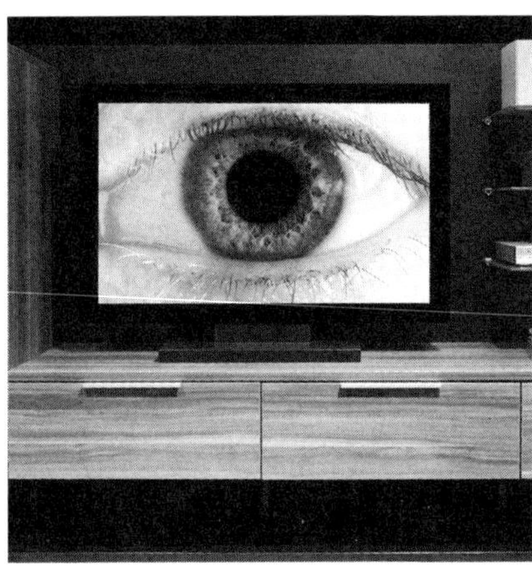

Might modern smart TVs have the ability to listen in on your conversations at home just as the author George Orwell once predicted they would?

Terms of Service. The aforementioned George Orwell would undoubtedly be spinning in his grave had he the opportunity to read the words. Samsung doesn't hide the startling facts; in fact, the company makes them acutely clear: "Please be aware that if your spoken words include personal or other sensitive information, that information will be among the data captured and transmitted to a third party through your use of Voice Recognition."

Never mind just a third party: with the current, sophisticated state of hacking, we could be talking about fourth, fifth, and sixth parties—in fact, endless numbers of parties who are secretly and carefully scrutinizing just about every word you utter while you sit in front of your TV. That we should even have gotten to this state is bad enough, but what comes next in this saga is even worse.

Clark Howard was not the only person concerned by all this outrageous spying in the one environment where we should not have jackbooted scum listening to us. Also hot on the trail of this story was Caleb Denison at *Digital Trends*. To a degree, his words play down the conspiracy angle. While he admits that the idea of having a television that, in effect, is

a spy in the home is "pretty nefarious," he also states that, in his opinion, much of what has been said about this particular issue has been "taken out of context." But is that really the case? Let's see.

On the matter of what Denison refers to as Samsung's "overly succinct description" of smart televisions, we are informed that when words and conversations may be recorded, we don't know for sure what exactly happens with them. Denison continues that the specific nature of the conditions under which the transfer of the data makes its way to a "third party"—which may be an ideal euphemism for an intelligence agency of the government—is not entirely clear.

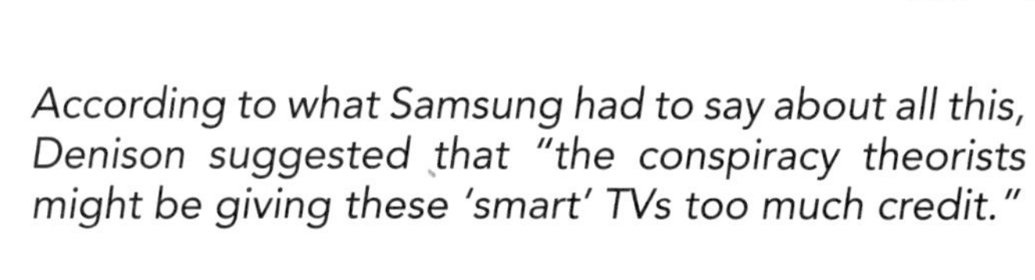

According to what Samsung had to say about all this, Denison suggested that "the conspiracy theorists might be giving these 'smart' TVs too much credit."

In light of all this, Denison decided to do the right thing and approach Samsung for the answers. According to what Samsung had to say about all this, Denison suggested that "the conspiracy theorists might be giving these 'smart' TVs too much credit." The writer's thoughts on this issue were prompted by the fact that, for the most part, smart TVs remain in the equivalent of a laptop's sleep mode. They correctly note that such televisions are designed and programmed to respond to specific words and statements made by people within the home in which the TVs can be found. We're talking about phrases like "Hi, TV" or "Hello, television." Denison at *Digital Trends* assures readers that without such phrases being specifically said, the TV does not—and cannot—take note of your spoken words and equally does not and cannot record and store any conversations that might conceivably be picked up in the home.

While this observation most assuredly played down the conspiratorial nature of this story—which, to this day, is showing no signs of going away at any time soon—there were other media outlets who were firmly of the opinion that the spies of the National Security Agency might soon be in the living rooms of all of us, albeit in a strange and stealthy way.

Writers for *Betanews*, for example, revealed their findings on this matter, specifically with regard to Samsung. The news was not good; at least, not for consumers. For the likes of the NSA, however, it was an absolute dream come true: "The company had publicly acknowledged that it was indeed logging users' activity and voice commands." *Betanews* also highlighted the following words from Samsung: "[T]hese functions are enabled only when users agree to the separate Samsung Privacy Policy and Terms of Use regarding this function when initially setting up the TV."

As *Betanews* dug further into the controversy, the publication noted the research in this field that had been undertaken by an English company, Pen Test Partners. The company's David Lodge and Ken Munro had gotten their hands on one of Samsung's smart TVs, chiefly to see if Samsung's claims were accurate or if we were all being deceived. The startling conclusion was that, no, Samsung, was not telling the complete story. In fact, far from it.

Despite Samsung's bold assertions to the contrary, Lodge and Munro were firmly able to demonstrate that the claims that all of the user data on users' smart TVs was completely encrypted and utterly safe from penetration were, frankly, complete and utter garbage. Secure was now insecure. The pair were able to prove that audio-based files, for example, were uploaded by Samsung's smart TVs in a wholly *un*encrypted fashion. This, of course, meant that any hacker with a high degree of smarts could access the material. And that doesn't just mean some amateur hacker; it also means those who may wish to exert a degree of control over us—such as the government and the intelligence community.

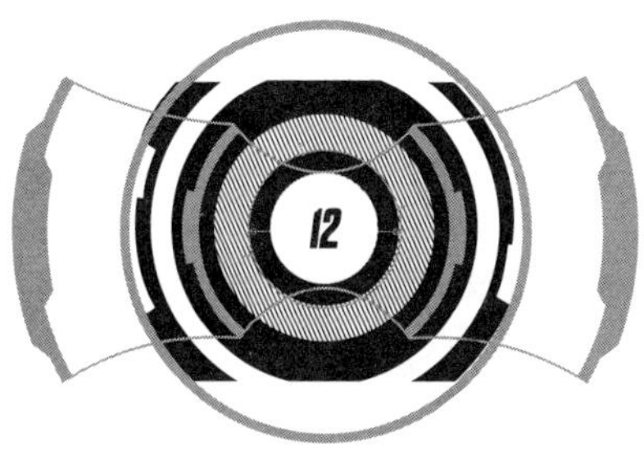

12 Robots in the Sky

There can be very few people who, today, have not heard of drones—or, to give them their more accurate title, unmanned aerial vehicles (UAVs). Many people own them and derive a great deal of entertainment from them. And there is nothing wrong with that! There is, however, a darker side to drones. It's a *much* darker side, related to the ways in which—when in the hands of government, the military, and the intelligence community—drones can be used to control us by spying on our every outdoor activity. Indeed, in recent years, the rise in the use of drones not just to spy on enemy nations or to seek out terrorists but to watch ordinary citizens, too, has increased massively.

The RAND Corporation explains: "An unmanned aerial vehicle (UAV) is an aircraft that carries no human pilot or passengers. UAVs—sometimes called 'drones'—can be fully or partially autonomous but are more often controlled remotely by a human pilot. RAND research has contributed to the public discussion on the use of drones for warfare and surveillance."

RAND notes that drones were initially utilized in warfare to provide tactical data but that today things are very different. Drones are playing significant roles in monitoring the likes of cell-phone towers and skyscrapers, primarily to ensure that there are no structural issues and, if there are, to help ensure that they can be repaired quickly. RAND also notes that drones are now being used increasingly in commercial ventures, such as movies and television shows. Such is the clamor for drones by both the government and the public that the Federal Aviation Administration has stated that, based on current trends, an astonishing three million drones will be in the skies by 2020. The FAA notes too that this figure is almost certainly destined to rise, particularly because of the huge plans that the likes of Google and Amazon have to employ drones in the workplace.

Most military UAVs have fixed wings, but there are exceptions such as the Navy's MQ-8B Fire Scout.

In other words, drones are everywhere. In some respects, that's good, particularly if the employment of such drones benefits us, the public. But what about when that groundbreaking technology is turned against us? That's when things get disturbing. We're talking about spies in the skies—here, there, and just about everywhere.

The Electronic Frontier Foundation (EFF), which has been at the forefront of warning people of the perils of drones becoming massive tools of control and surveillance, offers the following words: "Surveillance drones or unmanned aerial systems (UASs) raise significant issues for privacy and civil liberties. Drones are capable of highly advanced surveillance, and drones already in use by law enforcement can carry various types of equipment including live-feed video cameras, infrared cameras, heat sensors, and radar."

The EFF also makes an important observation that few are aware of: namely, that certain high-tech drones—primarily those of the military, the intelligence community, and worldwide governments—are able to stay aloft not just for hours but for days, endlessly spying, probing, and keeping watch on just about anything and anyone that their controllers decide needs to be watched. Drones are not just watching us, though, as the EFF has pointed out: drones can also be equipped with bullets and Taser-style technology, effectively turning them from spies to weapons.

One problem with all of this, the EFF correctly notes, is that while drone technology is advancing at a speedy and astonishing rate, the laws surrounding public privacy, including what can and cannot be done in the name of national security, are advancing at a far less speedy rate. For example, how many people are aware that drones of the government can monitor phone calls? Probably not many, but it doesn't end there: drones can access your messages on social media, read your texts, and check out your photos online. As for their ability to watch us intensively from the skies, the U.S. military is now using drones that are so advanced that they can read handwritten words from a height of around 20,000 feet.

Of course, when it comes to defending the free world from the actions of rogue nations and terrorists, the rise of the drone is not a bad thing, but like all technologies, drones can be used to not just spy on the enemy but also to intrude upon the lives of normal, everyday, law-abiding citizens. Why? Because they can. As a result of the crazed culture of surveillance—whether it is warranted or not—that reasoning now dominates much of today's society. Indeed, the picture is both grim and disturbing.

At the time of writing, plans are being initiated to extend the abilities of all government-owned drones to allow them to use infrared technology to monitor our towns and cities at night. The use of what is known as facial-recognition technology is now becoming a more and more relevant aspect of the drone programs of numerous governments all around the planet. Perhaps the most unsettling aspect of all this is the issue of what we might call minidrones. As incredible as it might sound, the highly

As incredible as it might sound, the highly visible drones of today are quickly going the way of the dinosaurs and the dodo. They are being replaced by drones that are, amazingly, only inches in size.

visible drones of today are quickly going the way of the dinosaurs and the dodo. They are being replaced by drones that are, amazingly, only inches in size.

On the matter of how the use of drones is taking us down the path toward a definitive *1984* world, one only has to take a look at certain events that were revealed in April 2014. Of the affair in question, *The Atlantic* states: "In a secret test of mass surveillance technology, the Los Angeles County Sheriff's Department sent a civilian aircraft over Compton, California, capturing high-resolution video of everything that happened inside that 10-square-mile municipality. Compton residents weren't told about the spying, which happened in 2012."

The Center for Investigative Reporting was told by Ross McNutt of Persistence Surveillance Systems (PSS): "We literally watched all of Compton during the times that we were flying, so we could zoom in anywhere within the city of Compton and follow cars and see people." PSS is currently making major headway in terms of selling such technology to police forces. PSS's drones can fly the skies for more than five hours, and they have TiVo-style technology that not only allows drones to watch us in real-time but also to rewind to view recordings made minutes or hours earlier.

Again, when used for positive, nonintrusive reasons, this is all good. The problem comes when the technology is abused. On this matter, *The Atlantic* also said something extremely thought-provoking: "A sergeant in the L.A. County Sheriff's Department compared the experiment to Big Brother, even though he went ahead with it willingly. Is your city next?"

And that's the problem: when agencies of government know they are pushing us toward a domain filled with George Orwell's worst nightmares, and when the employees of those same agencies just go along with it and fail to speak out about it, it's no wonder we are spiraling into a surveillance state.

Moving on from the United States, we note that the last few years, particularly in the wake of the Edward Snowden affair, the U.K.'s *Guardian* newspaper has been at the forefront of a concerted effort to wake people up to the intrusions in our society that threaten our privacy and our civil liberties. In 2010, in a chilling and eye-opening article, the *Guardian* noted: "More and more police forces and government agencies are exploring the potential of unmanned drones for covert aerial surveillance, security, or emergency operations across the UK, the *Guardian* has learned."

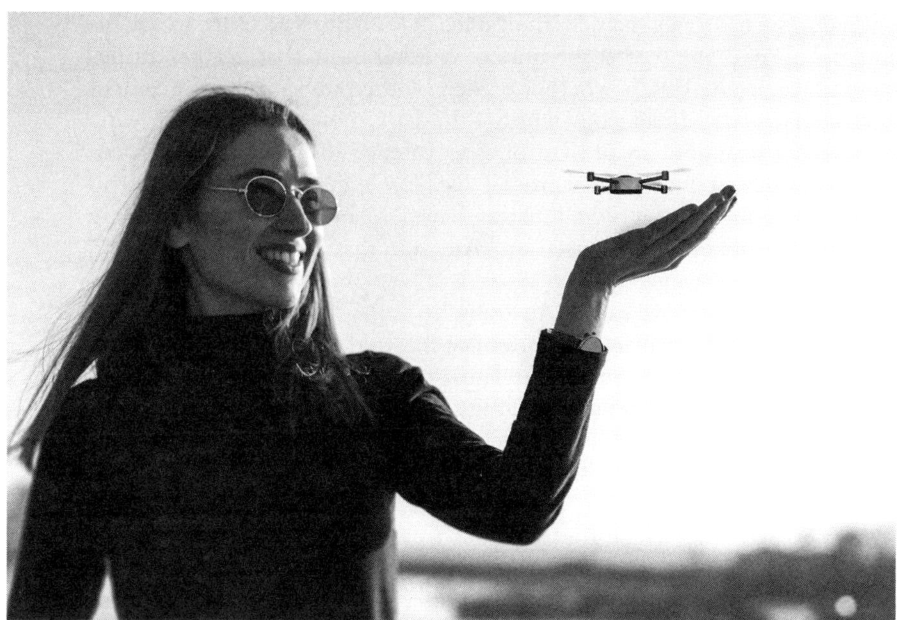

Everything in the world of high tech is getting smaller, including drones.

According to the *Guardian*, one of the departments of the British government that is at the forefront of keeping the population of the United Kingdom under surveillance is SOCA, the Serious Organized Crime Agency. SOCA has a close working relationship with MI5 (Military Intelligence, Section 5), which is the U.K.'s equivalent of the United States' FBI (Federal Bureau of Investigation). Government documentation secured by investigative journalists of the *Guardian* revealed that the growth of the use of drones in the U.K. by regional police forces is growing at an extremely rapid rate. And it's not just the *Guardian* that has reported on such issues in the United Kingdom.

The BBC, in April 2015, told its readers: "Police guarding London airports will start using drones for surveillance following a review by counter-terrorism officers. An 18-month analysis by the National Counter Terrorism Policing Headquarters, which helps develop police policy, found the technology could be 'transformative.' Privacy campaigners said they were concerned about the plans. Police are also to take over investigations into drone misuse."

It's decidedly disturbing that the very people who, in the U.K., are tasked with investigating the misuse of drones are the same people who are using such drones, namely the police. The BBC had more to say on this issue. Revealing that at the time of writing more than 50,000 drones were in use in the U.K.—whether owned by the public, law-enforcement personnel, or the intelligence world—the BBC noted how this growing increase in the employment of drones threatened the privacy of the people of the U.K. It quoted the words of *Big Brother Watch*'s Emma Carr, who said about drones that it is vital that we have a fuller grasp of "how they're used and who they're being used by." Amen to that.

Now let us move on to Australia. In 2014, Australia's ABC News reported: "The Australian Federal Police has told a federal parliamentary inquiry that drones are among the latest tools it is using in major crime investigations. A parliamentary committee has begun examining the growing use and popularity of drone technology, as well as the privacy concerns it raises for Australians."

The extent to which the AFP was using drones, and the nature of the technology involved, became very clear in 2013. That was when the AFP used drone-based technology to search for the remains of an anti-drugs proponent in New South Wales. Mark Harrison, the commander of the AFP, told the Australian Press the following: "We assisted in the imagery and the recording of that search and excavation and the UAV provided a different and unique perspective to aid and assist in that process." Notably, this was the very first occasion on which the AFP acknowledged it had a drone program.

As is the case in the United States and the United Kingdom, the citizens of Australia have their own deep concerns about how drones may be misused in the name of what passes for national security. Certainly, the provisions of the Australian Freedom of Information Act have demonstrated that more and more people have lodged complaints with the world of law enforcement, chiefly to express their outrage at what they see as flagrant disregard for personal privacy. Much of this was revealed by Australia's Office of the Privacy Commissioner. Timothy Pilgram, the commissioner himself, admitted that "there are cases where people have woken up in the morning, pulled their curtains open and there's been a drone hovering outside. It's starting to suggest to us that there's a point in time at which we need to sit back and say: 'Do we have the right laws in place to make sure that we can regulate to the best extent possible how these devices can be used?' That is the challenge now facing lawmakers in Federal Parliament and across Australia."

In Australia, at least, there is some degree of oversight: the police, the military, and the intelligence community are now required to secure a warrant before any kind of drone-driven surveillance of citizens and residents can proceed—which is very encouraging. We can only hope that such rules and regulations are being adhered to.

Now it's time to move onto the issue of how drones are practically mutating into the kinds of things one would expect to see in the realm of science fiction. It's an issue that revolves around what are now widely referred to as "insect drones."

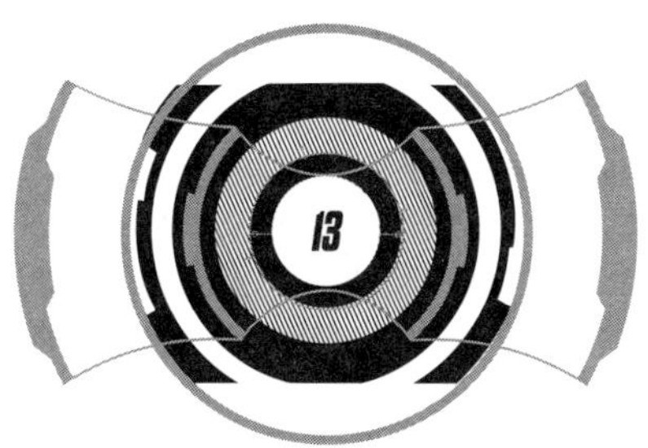

The Rise of the Insect Robots

The phenomenon of the insect drone is a growing one—although, ironically, as the name suggests, they are just about the smallest drones conceivable. Back in December 2014, Patrick Tucker, writing at Defense One, stated that the Defense Advanced Projects Research Agency (DARPA) "put out a broad agency announcement this week seeking software solutions to help small drones fly better in tight enclosed environments."

DARPA's statement concerning its Fast Lightweight Autonomy program, as it is called, said in part that the operation was based around "creating a new class of algorithms to enable small, unmanned aerial vehicles to quickly navigate a labyrinth of rooms, stairways and corridors or other obstacle-filled environments without a remote pilot." Tucker noted that while "urban disaster" was an environment in which small, insect-like drones could be used in a positive fashion, it was also now possible for DARPA and the agencies working with it to use these mini-drones to "fly independently into rooms, find a perch, and serve as a fly on the wall in a very real (but robotic) sense."

Moving on to the summer of 2016, that was when the United Kingdom's *Daily Mail* newspaper was hot on the trail of the expanding phenomenon of insect drones. The newspaper's defense correspondent, Larisa Brown, wrote that a new, small device, "modelled on an insect," was fast becoming the U.K.'s "latest weapon against terror." Brown added: "The Dragonfly drone—which can fit in the palm of a hand—will spy on enemy positions and gather intelligence for the military and British agents.

It is inspired by the biology of a dragonfly, with four flapping wings and four legs to enable it to fly through the air seamlessly and perch on a windowsill to spy on terrorists."

The astute readers of the *Daily Mail* aired their grievances and concerns in the "Comments" section of the newspaper. In the words of one such person: "Only the blind would fail to see something that size flying into a room. More likely to be used by the establishment so that Big Brother can keep a closer eye on us."

The newspaper's defense correspondent, Larisa Brown, wrote that a new, small device, "modelled on an insect," was fast becoming the U.K.'s "latest weapon against terror."

Another stated: "It scares me that they have technology like this but does not surprise me. It is only a matter of time before they will be spying on every one of us just like George Orwell predicted."

The final words on this sinister phenomenon of the rise of the insect drones goes to Paul Joseph Watson at the *Infowars* website. He states: "Harvard Professor Margo Seltzer warned that miniature mosquito drones will one day forcibly extract your DNA on behalf of the government and insurance companies as she told elitists at the World Economic Forum in Davos that privacy was dead."

Despite the controversial nature of this statement from Professor Seltzer, it's important to note she is no wild-eyed, paranoid conspiracy theorist. Indeed, she is nothing less than a professor in computer science at Harvard. She additionally told the audience at the World Economic Forum, "It's not whether this is going to happen, it's already happening. We live in a surveillance state today.

We certainly do. And as the control of the world's populations grows, there is very little doubt that insect drones will play a leading role in that control—unless we all band together to swat these annoying things, just like we would a real insect.

Now consider the following from the BBC: "When Storm Ciara swept across the UK in February, Alex Caccia was strolling on Oxford's Port Meadow watching birds take to the air. He marveled at their indifference to high winds: 'While airliners were grounded by the weather the birds couldn't care less!' It was more than just a passing thought for Mr Caccia, who is the chief executive of Animal Dynamics, a technology start-up applying lessons from wildlife to drone design.

"Formed in 2015 to pursue the science known as biomechanics, his company already has two drones to show for an intimate study of bird and insect life. One takes inspiration from a dragonfly and has attracted funding from the military. Its four wings make it steady in high winds that would defeat existing miniature spy drones. Known as Skeeter, the secretive project has cracked the challenge of using flapping wings to power a drone. While wings are more efficient than a propeller and allow a dragonfly to hover in the face of strong gusts, they are almost impossible for human engineers to emulate."

Rod Vaughan at New Zealand's *National Business Review* has his own opinions on the world of the robot insect and its growing role in keeping watch on us—potentially all of us, one day:

> In the US there are claims the federal government has been covertly using such drones for several years, employing them to monitor anti-war protests and the like. In 2007 university student Vanessa Alarcon, who was involved in an anti-war rally, told the *Washington Post*: "I look up and I'm like, 'what the hell is that?' They looked like dragonflies or little helicopters, but I mean, those are not insects." Lawyer Bernard Crane, who was at the same event, reportedly said he had "never seen anything like it" in his life. "They were large for dragonflies and I thought, 'is that mechanical or is that alive?'"

Then, there are the words from Daniel Ackerman at *MIT News*: "If you've ever swatted a mosquito away from your face, only to have it return again (and again and again), you know that insects can be remarkably acrobatic and resilient in flight. Those traits help them navigate the aerial world, with all of its wind gusts, obstacles, and general uncertainty. Such traits are also hard to build into flying robots, but MIT Assistant Professor Kevin Yufeng Chen has built a system that approaches insects' agility.

"Chen, a member of the Department of Electrical Engineering and Computer Science and the Research Laboratory of Electronics, has developed insect-sized drones with unprecedented dexterity and resilience. The aerial robots are powered by a new class of soft actuator, which allows them to withstand the physical travails of real-world flight. Chen hopes

Bee-sized drones our now possible. They could be used for surveillance and spying on perceived foreign and domestic threats.

the robots could one day aid humans by pollinating crops or performing machinery inspections in cramped spaces."

At *Robotics Tomorrow*, Len Calderone provides us with an example of how "a police officer can command the insect drone to 'hover' or 'follow me.' If a foot chase develops during a traffic stop, the officer can command the drone to follow the suspect so that the officer can attempt to cut the suspect off. If there are two or more suspects, the officer can command the drone to follow one of the suspects, while he gives chase to the other.

"Suppose you look forward to the latest iPhone but there is a long line. You want to know how many people are ahead of you to determine if the wait is worth it. You can simply say 'hover' to the insect drone and the drone can take up a position at a predetermined distance above the user. Maybe you can just wait for that new phone. This same drone can be used to find someone in a crowd.

"A product of the U.S. Army, the robo-fly, will be used for spying. Speeding through dangerous indoor spaces at 65 feet per second, the insect drone would allow reconnaissance to be conducted with no risk to human life. The wings of the robo-fly are made of zirconium titanate (PZT), a material that flaps and bends when voltage is applied. The British military has a spy drone that is the size of a dragonfly and can fit in the palm of a soldier's hand. This tiny drone has four wings and four legs so that it can fly seamlessly through the air and perch on a windowsill like a dragonfly. It can fly into a room full of enemy soldiers and report back."

The American Association for the Advancement of Science gives us this: "A new timeline for insects shows that the creatures first evolved 479 million years ago—earlier than previously suspected—and that their appearance coincided with Earth's first land plants. The revised insect tree of life, also called a phylogeny, is helping scientists answer some of the longstanding questions regarding the origin of these highly diverse and tremendously important arthropods. But constructing the phylogeny was no picnic. Insects are by far the most species-rich group of animals on the planet, and the researchers involved had to use several supercomputers to process all of their genetic data."

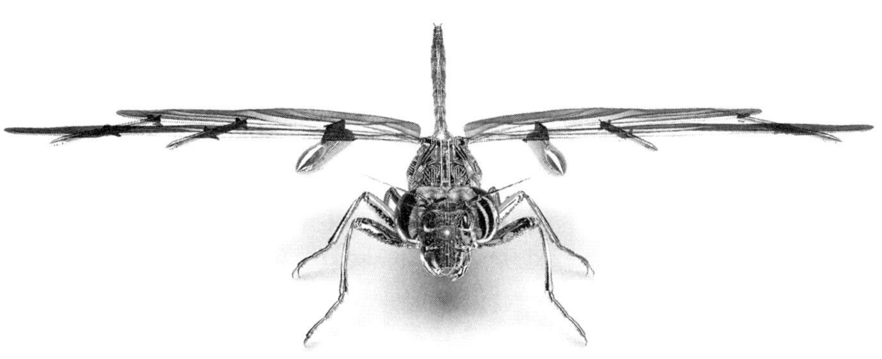

The CIA was trying to create a dragonfly robot as long ago as the 1970s. Fifty years later, tech companies are still working on this challenging goal.

Army Technology expands on all this: "It may sound like the stuff of science fiction, but it has been the goal of real-life military research for decades, arguably beginning in earnest in the 1970s with the CIA's attempt to perfect a gas-driven robot dragonfly—the 'insectothopter'—to help its covert intelligence gathering operations. Although that project ultimately came to nothing, forty years on new developments in microelectromechanical systems (MEMS) and DARPA's recently announced Fast Lightweight Autonomy (FLA) program mean that robotic spy-flies are now closer than ever to becoming a reality. Nevertheless, there are still some challenges ahead.

"'Scientists and engineers have a reasonable understanding of the aerodynamics of insect-like and insect-scale flight; however, we're still significantly limited by the power, sensing and computing resources that can be put on such small aircraft,' says Dr Larry Matthies, senior research scientist and supervisor of the Computer Vision Group at the Jet Propulsion Laboratory in Pasadena, US. 'There are also limitations in the performance of actuators and drivetrains to move the wings efficiently.'"

In 2006, the Pentagon's DARPA asked scientists to create cyborg insects. With DARPA's support, researchers at the University of California–Berkeley successfully created a cyborg beetle whose movements they could remotely control. They reported their results in *Frontiers in Integrative Neuroscience* in October 2009. "Berkeley scientists appear to have demonstrated an impressive degree of control over their insect's flight; they report being able to use an implant for neural stimulation of the beetle's brain to start, stop, and control the insect in flight," reported *Wired* the month these findings came out. "They could even command turns by stimulating the basalar muscles."

"We demonstrated the remote control of insects in free flight via an implantable radio-equipped miniature neural stimulating system," the researchers wrote at *Frontiers in Integrative Neuroscience* "The pronotum mounted system consisted of neural stimulators, muscular stimulators, a radio transceiver–equipped microcontroller and a microbattery. Flight initiation, cessation and elevation control were accomplished through neural stimulus of the brain which elicited, suppressed or modulated wing oscillation. Turns were triggered through the direct muscular stimulus of either of the basalar muscles. We characterized the response times, success rates, and free-flight trajectories elicited by our neural control systems in remotely controlled beetles. We believe this type of technology will open the door to in-flight perturbation and recording of insect flight responses."

Eyes in the skies? There's absolutely no doubt about that.

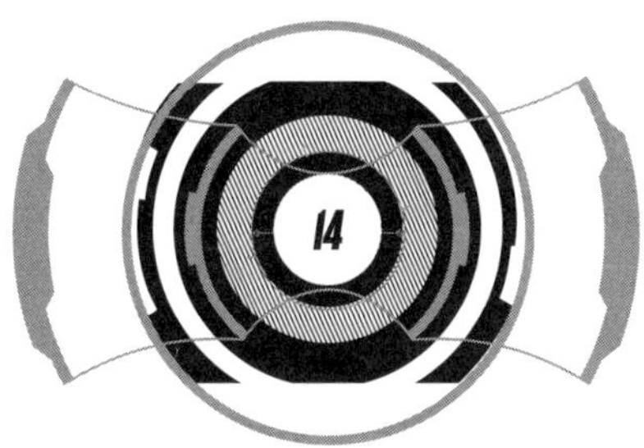

Robots Doing What We Can Do

The A.I. Research and Advisory Company provides us a fascinating look at the way robots are proliferating in the family home: "The International Federation of Robotics reports that the U.S. service robot industry, which includes both industrial and domestic sectors, is a $5.2 billion market. It also estimates that home robots or domestic robots will contribute $11 billion in revenue by 2020. Home robots have existed since the 1990s. An early example of this is the 2001 Electrolux robot vacuum cleaner. They are currently used for helping humans with many kinds of domestic chores."

Consider also the words of Tim Hornyak in *Digital Arts* in May 2014: "While debate on military robots heated up this month thanks to UN [United Nations] talks about the development of lethal autonomous robots—and military robots are evolving quickly thanks to defense budgets—household robots remain far from ubiquitous. More than a half-century after the world's first industrial robot, Unimate, began work at a General Motors plant, most commercial robots still work in factories. The ones that are in households, such as the roughly 10 million robot vacuum cleaners led by iRobot's Roomba, have usually been limited to performing one task only, like sucking up dirt. Computers and robots can beat us at dedicated tasks like chess or painting cars, though humans still have a massive intelligence advantage in terms of general knowledge. That's a good thing if you fear a robot uprising. Not so great if you're waiting for that perfect humanoid robot servant from science fiction films, like a C-3PO."

Did You Know Homes points out the many industries where robots are employed: "*Restaurants*: Robots are used in kitchens to prepare food and chop vegetables. This technology is used in Japan for making sushi and chopping vegetables. Robots also being experimented with to work as receptionists, cleaners in many restaurants and homes. *Crime Fighting*: Robots are used in police forces to check buildings to pinpoint the location of criminals or dangerous devices. Remotely controlled robots are used to check cars for bombs which they can also be programmed to disarm. *Medicine*: Robots can be used in the hospital to distribute medication to patients. They can be programmed to interface with intelligent hospital elevators to reach any floor and much more. *Education*: Robots are being utilized as teacher's assistants and designed to care for children and there are also robotic toys for children."

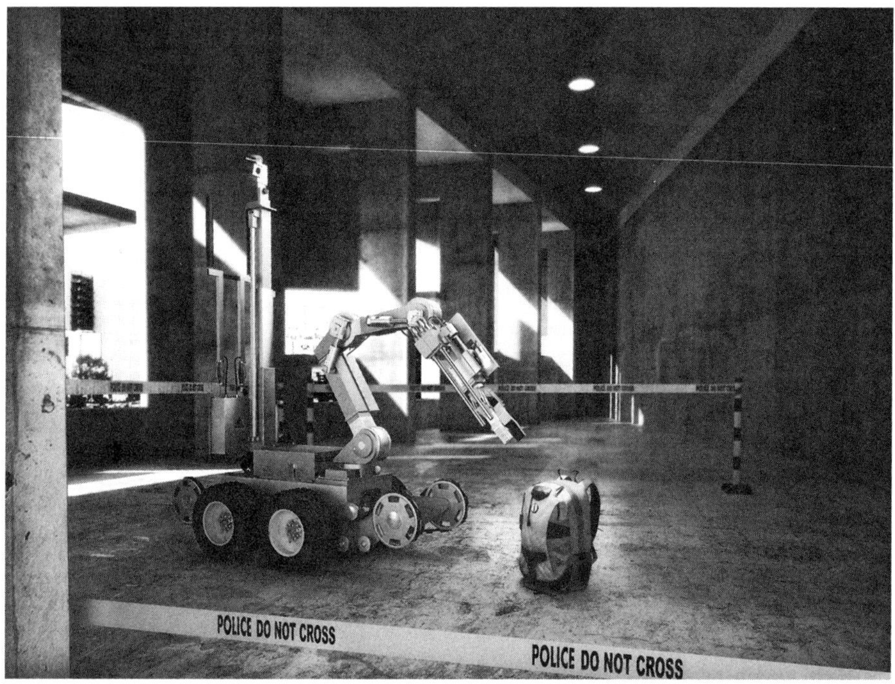

A police bomb squad robot is run through a test to investigate a suspicious carrying case. Robots can be highly useful in performing dangerous tasks such as this one.

Now we head over to Brett Tingley, a feature writer at the popular website *Mysterious Universe*: "I don't know about you, but I can't wait until robots take their rightful place in the world. The science fiction of our childhoods promised us so many wondrous things like flying cars, pneumatic tube travel, and off-world colonization, but most of those are still so disappointingly far away. Yeah, we've got real-life *Star Trek* communicators in our pockets and a few crummy space stations in orbit, but so what? I want walking, talking, beeping-booping robots with square pupils, damnit. With a little luck (and funding for university robotics departments), that will soon be a reality.

Robots have already taken over the manufacturing industry and are on their way to doing the same with long-distance hauling, taxi driving, combat roles in the military, pizza delivery, surgery, and even sex workers.

"Robots have already taken over the manufacturing industry and are on their way to doing the same with long-distance hauling, taxi driving, combat roles in the military, pizza delivery, surgery, and even sex workers. But me, I won't be happy until an android with an over-the-top upper-crust British accent hands me my coffee at Starbucks or accompanies me to translate the binary language of moisture vaporators. That might be closer than we think, though. Thanks to recent advances in robotics and robot–human relations, the world was made a little weirder as the first robot was just granted a national citizenship."

Communing with Robots and Craving for Immortality

The year 1987 was one of the most important in the field of UFO research. That was the year in which Whitley Strieber's book *Communion* was published. The alien abduction–themed book became a huge bestseller—even a *New York Times* bestseller, no less. Strieber's book was an important one because it didn't follow the same pathway that so many other abduction books did. Strieber postulated that the black-eyed, dwarfish creatures known as "the Greys" are far from being just entities from another world who have come to visit us. Strieber's work suggested something far more than that—something that has deep relevance to the world of robots.

Some people were excited and intrigued by the story that Strieber told. Some were puzzled and others appalled by the fact that *Communion* brought together matters relative to magic, the human soul, immortality, and life after death, all in one big mix. Indeed, Strieber put forth the distinct possibility that the Greys have the ability to operate in both a physical state and a spiritual or soul-based one. As for the physical form, Strieber came up with a fascinating theory based on his own experiences with the Greys. He strongly suspected that when not in spirit or soul form, the Greys had the ability to be "downloaded" into physical form. Furthermore, in that particular form, they were far more like biological robots than the aliens that the UFO research community embraced—in fact, not unlike those spoken about by Nigel Kerner and Philip Corso, whom we discuss in the chapter "Roswell, Robots, and UFOs."

Futurologist Mac Tonnies went down that pathway, too. He said: "Given the vast number of out-of-body and near-death experiences, I find it difficult to reject the prospect of 'nonlocal' consciousness; perhaps a sufficiently advanced technology can manipulate the 'soul' as easily as we splice genes or mix chemicals in test tubes. If so, encounters with 'extraterrestrials' may help provide a working knowledge of how to modify and transfer consciousness—abilities that seem remote to the current terrestrial state-of-the-art, but may prove invaluable in a future where telepresence and virtual reality are integral to communication. Already, the capabilities of brain-machine interfaces are tantalizingly like the popular perception of telepathy, often thought of in strictly 'paranormal' or even 'magical' terms."

It should be noted that this is closely related to a philosophy known as transhumanism. Basically, it's a case of being in a post-human rather than present-human state—or even post-alien rather than alien. There is even a political party known as the Transhumanist Party. With transhumanism, we're talking about us, as a species, becoming something amazing—a new type of human. Becoming *post*-human and, potentially, immortal. We're even talking about the merging of man and highly advanced technology in ways that, right now, are difficult to understand but could be a viable way to ensure we have far more than our measly 80 years, or thereabouts, and instead may even live forever. It's hard today to get one's head around the concept, but that may not be the case for people a century away. For them, immortality may just be an everyday thing. The man who has flown the flag for transhumanism is Zoltan Istvan, who is determined to become both immortal and president of the United States—in which particular order hasn't been confirmed yet.

Evolutionary biologist and eugenicist Sir Julian Huxley coined the term "transhumanism" in 1951. He believed that human beings should strive to improve themselves as a species.

The BBC says of this topic: "Transhumanists dream of achieving immortality and physical perfection through futuristic technologies like mind uploading, cyborg body augmentation, and genetic manipulation; they want us to evolve into a race of post-human super-beings. While the other presidential candidates are claiming they know best how to deal with Iran, the economy, and immigration, Istvan is trying to appeal to the US electorate with more ambitious goals. He wants to eradicate death, and for everyone in America to live forever." The *Telegraph* said, in 2014, that those within the Transhumanist Party are "largely rich Californians, technology geeks and scientists (sometimes all three)."

Istvan has said that all of this "literally means *beyond human*. Transhumanists consist of life extensionists, techno-optimists, Singularitarians, biohackers, roboticists, AI proponents, and futurists who embrace radical science and technology to improve the human condition. The most important aim for many transhumanists is to overcome human mortality, a goal some believe is achievable by 2045."

There are those who are concerned not about transhumanism *per se* but about the developing of a nonhuman state. In other words, becoming immortal might just wipe out what it is that makes us human. A man named Nick Bostrom is concerned by at least some of the work being done in the world of transhumanism. Bostrom has an impressive bio. This is just a part of it:

"Nick Bostrom is a Swedish-born philosopher and polymath with a background in theoretical physics, computational neuroscience, logic, and artificial intelligence, as well as philosophy. He is a professor at Oxford University, where he leads the Future of Humanity Institute as its founding director. (The FHI is a multidisciplinary university research center; it is also home to the Center for the Governance of Artificial Intelligence and to teams working on AI safety, biosecurity, macrostrategy, and various other technology or foundational questions.) He is the author of some 200 publications, including *Anthropic Bias* (2002), *Global Catastrophic Risks*

A professor at the University of Oxford, Sweden-born philosopher Nick Bostrom is founding director of the Future of Humanity Institute at Oxford and also the founder of the Oxford Martin Program on the Impacts of Future Technology.

(2008), *Human Enhancement* (2009), and *Superintelligence: Paths, Dangers, Strategies* (2014), a *New York Times* bestseller which helped spark a global conversation about artificial intelligence. Bostrom's widely influential work, which traverses philosophy, science, ethics, and technology, has illuminated the links between our present actions and long-term global outcomes, thereby casting a new light on the human condition."

Bostrom says of transhumanism: "It is difficult for us to imagine what it would be like to be a posthuman person. Posthumans may have experiences and concerns that we cannot fathom, thoughts that cannot fit into the three-pound lumps of neural tissue that we use for thinking. Some posthumans may find it advantageous to jettison their bodies altogether and live as information patterns on vast super-fast computer networks. Their minds may be not only more powerful than ours but may also employ different cognitive architectures or include new sensory modalities that enable greater participation in their virtual reality settings. Posthuman minds might be able to share memories and experiences directly, greatly increasing the efficiency, quality, and modes in which posthumans could communicate with each other. The boundaries between posthuman minds may not be as sharply defined as those between humans."

Bostrom gives us a great deal of food for thought when it comes to transhumanism.

But transhumanism isn't the only way we might achieve immortality. In the latter part of 2015, astounding information came out of Russia's State University, Moscow. The man who created this wealth of controversy was one Anatoli Brouchkov. Around 2013, Brouchkov injected himself with a certain form of bacteria. The specific strain was nothing less than a 3.5-million-year-old strain of *Bacillus F* from Siberia.

Writer Paul Seaburn takes the story further, writing in an article dated October 2, 2015: "*Bacillus F* was discovered in 2009 in permafrost located at a site called Mammoth Mountain in the Sakha Republic—the largest and coldest region of Siberia. Since then, scientists have been studying it and last month announced that they had decoded its DNA. That was nothing compared to the revelation this week that Dr. Anatoli Brouchkov, head of the Geocryology Department, Moscow State University, injected himself with the bacteria two years ago. Is he mad?"

First, Brouchkov and other scientists had tested the bacteria on mice, plants, and fruit flies. Among other results, elderly mice were once again able to reproduce. "Those results convinced Brouchkov to try some inactivated bacterial culture on himself," writes Seaburn. "After two years, he's still alive. That's a good sign, but a slightly better one is that he says he hasn't had the flu since becoming a human guinea pig. Brouchkov is obviously hoping for more than just flu season relief. The Yakut people native to the Mammoth Mountain area appear to live longer than average and he believes it's due to the bacteria entering their drinking water when the permafrost melts. He also feels it has given him more energy and says this may be an indication of its 'bacteria of youth' potential."

Of course, things are very much still in the early stages, so it's still a case of wait and see. While this isn't strictly related to transhumanism, it demonstrates how people are now doing more and more to try to keep the Grim Reaper away. Marios Kyriazis, a medical doctor, states in "Extreme Lifespans through Perpetual-Equalizing Interventions (ELPIs)": "It is nonsensical and counter-intuitive to believe that complex life was created only to end after a set period of time. An intelligent, complex being should be able to live indefinitely, or to put it in another way, it should not be allowed to die through ageing."

We haven't quite finished: Meet Ray Kurzweil. Let's first take a look at his background: "Ray Kurzweil is one of the world's leading inventors, thinkers, and futurists, with a thirty-year track record of accurate predictions. Called 'the restless genius' by *The Wall Street Journal* and 'the ultimate thinking machine' by *Forbes* magazine, Kurzweil was selected as one of the top entrepreneurs by *Inc.* magazine, which described him as the 'rightful heir to Thomas Edison.' PBS selected him as one of the 'sixteen revolutionaries who made America.'

"Kurzweil was the principal inventor of the first CCD flat-bed scanner, the first omni-font optical character recognition, the first print-to-speech reading machine for the blind, the first text-to-speech synthesizer, the first music synthesizer capable of recreating the grand piano and other orchestral instruments, and the first commercially marketed large-vocabulary speech recognition system."

The recipient of the 1999 National Medal of Technology and Innovation, Ray Kurzweil is a brilliant inventor and futurist who has been called the rightful heir of Thomas Edison. The author of works on transhumanism and AI, he believes that things like biotech and nanotech could make humans immortal.

At the Global Futures 2014 International Congress held in New York in 2013, Kurzweil said that as the years and centuries pass, the likelihood is that we will come to a point where "the non-biological part dominates and the biological part is not important anymore. In fact, the non-biological part—the machine part—will be so powerful it can completely model and understand the biological part. So even if that biological part went away it wouldn't make any difference."

It may well be the case that Kurzweil's words have a bearing on those biological robots that Philip Corso claimed to have seen at Roswell, New Mexico, in 1947, the strange "Grey" aliens that Whitley Strieber described, and the Women in Black, the Black-Eyed Children, and the Men in Black.

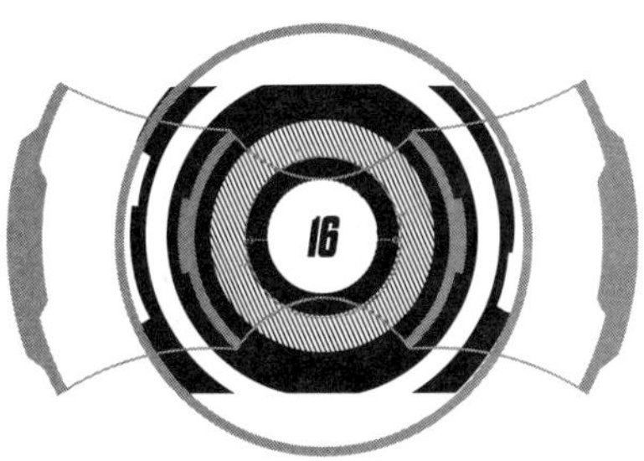

When a Robot Made the World Stand Still

Now we come to what is without doubt the defining decade in relation to all things robot: the 1950s. We're not talking about the world of fact, though, but the domain of fiction—specifically, one of the most famous sci-fi movies of all time, 1951's *The Day the Earth Stood Still.* Not only did it amaze and entertain audiences around the world, but it also brought to sci-fi fans a new fascination for all things of a robotic nature. In the wake of the movie, many tried to follow in its footprints. Most failed. As for the theme of the story, it was unlike most of what had come before. This was a movie designed to have its audience think and to show how we, the human race, were the most dangerous creatures on planet Earth. It was also meant to draw parallels between the world of on-screen fantasy and

A science fiction classic, 1951's The Day the Earth Stood Still *is about an alien visiting Earth to try to help the human race. He is accompanied by an imposing robot that protects the alien from attack.*

the growing threat of the Cold War. There's a fascinating allegory to the movie, too. With that said, let's take a look at the story.

Made by Twentieth Century-Fox, the movie starred Michael Rennie (who, before *The Day the Earth Stood Still*, appeared in 1945's *Caesar and Cleopatra*, and after it, in 1960, appeared in *The Lost World*). Rennie played a very human-looking alien with the name of Klaatu. Interestingly, there was no love interest or romance in the movie, an approach hardly ever seen in the sci-fi of the movies of the early 1950s. That's not to say there wasn't a leading lady. There was: Helen Benson, played by Patricia Neal. Notably, her character was not one driven by screams and fainting in every other shot, as in so many productions of its era. There was also the actor Sam Jaffe, playing an Albert Einstein–type character who comes to realize that the human race is not the most important species in the universe, nor is it the most scientifically advanced. Helen's son, Bobby Benson (played by Billy Gray), also had an integral role in the movie. Finally, there was Lock Martin, the man who played Gort. For those who may not know, Gort was arguably the star of *The Day the Earth Stood Still* even though he (or it) was a robot—a giant-sized robot.

We have to take note of the fact that it was in 1947 when the "flying saucer" rage began. Four years later, flying saucers were still big pulls in the theaters. Hollywood knew that. However, the crew—Robert Wise, the director; Julian Blaustein, the producer; and Edmund H. North, who handled the screenplay, based on Harry Bates's story "Farewell to the Master"—wanted something a bit different. Hollywood produced it.

As the movie begins, we see a definitive large flying saucer soaring across the skies of the world, eventually coming to land in Washington, D.C. People everywhere are amazed and frightened, particularly when a "doorway" in the craft opens and a humanoid figure appears. Things go deadly quiet. Naturally, the military moves in and gets a bit antsy, to say the least: suddenly, one of the contingent of soldiers in the area fires his gun, injuring what turns out to be an alien that looks just like us. Indeed, there is no green skin, bulging eyes, or

Gort, the robot companion to Klaatu in The Day the Earth Stood Still, *served as his master's protector.*

claws. And so on. If not for the spacesuit the being wears, he would be able to move among us with ease. Later in the movie, he does exactly that.

We learn that the name of the alien is Klaatu. His robot sidekick, Gort, is an ominous, towering robot that may look like a clunking, slow creation but actually has incredible destructive power. As the story expands, it becomes clear that Gort has the ability to wipe out not just the human race but the whole planet. Klaatu says: "For our policemen, we created a race of robots. Their function is to patrol the planets in spaceships like this one and preserve the peace. In matters of aggression, we have given them absolute power over us. This power cannot be revoked. At the first sign of violence, they act automatically against the aggressor. The penalty for provoking their action is too terrible to risk. The result is, we live in peace, without arms or armies, secure in the knowledge that we are free from aggression and war, free to pursue more profitable enterprises. Now, we do not pretend to have achieved perfection, but we do have a system, and it works."

As for Klaatu's home planet, we're never told the actual world from which the man from the stars comes. After being injured by the gun-happy soldier, Klaatu is taken to Walter Reed Army Hospital. When it becomes clear to Klaatu that U.S. authorities will not let him leave, he escapes, making his way to a boarding house and using the name of "Mr. Carpenter."

It doesn't take long before Bobby, his mother, and Helen's quasi-boyfriend, Tom Stevens—played by Hugh Marlowe—learn who Mr. Carpenter really is. Klaatu warns the world of the incredible power that the practically invincible Gort can unleash on the planet, but not before he is killed and resurrected. We're talking about the end of life on Earth. As Klaatu says: "Your choice is simple: join us and live in peace, or pursue your present course and face obliteration. We shall be waiting for your answer." Klaatu and mighty Gort take to the skies, while the people of Earth—particularly the world leaders—ponder on the future. If, that is, they are allowed to have a future.

There's another angle to all of this: a religious one. *Bright Lights Film Journal* explains what it calls the "Christian analogy" in *The Day the Earth Stood Still*: "We have free will and can decide to ignore the superior beings' instruction. However, to ignore it means we are condemned to a hellish end." Furthering the analogy, *Bright Lights Film Journal* adds: "The name 'Carpenter' refers to the Son of God's earthly vocation.... He emerges—resurrected—from the ship and delivers a harsh message before his ascension to the heavens.... Klaatu/Carpenter's gathering of scientists recalls the young Jesus talking to the rabbis in the temple."

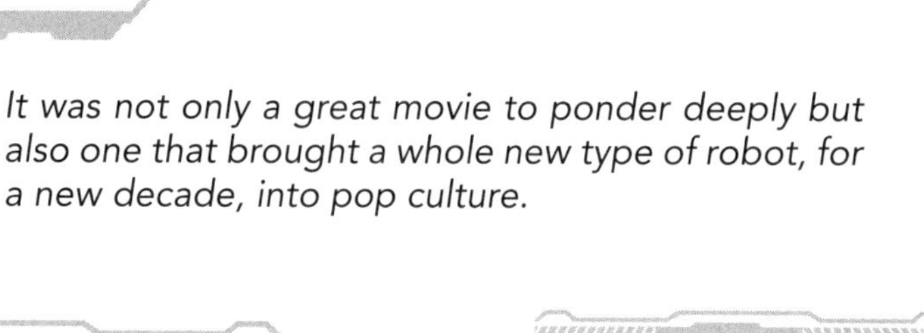
It was not only a great movie to ponder deeply but also one that brought a whole new type of robot, for a new decade, into pop culture.

It was not only a great movie to ponder deeply but also one that brought a whole new type of robot, for a new decade, into pop culture. It should be noted that the theme of *The Day the Earth Stood Still* inspired a group of predominantly American figures who became known in the UFO research arena as "contactees." They claimed face-to-face contact with Klaatu-like aliens and who warned of nuclear war and irreversible ecological danger. And they looked very much like Klaatu: enigmatic and wise and, on top of that, somewhat bullying in tone and action. One only has to read the following to see how *The Day the Earth Stood Still* drove the flying saucer community.

The contactees were people who, chiefly in the 1950s, claimed encounters with very human-looking aliens. Those same aliens demanded that we lay down our atomic weapons. Contactees were those controversial characters who claimed close encounters with very human-looking aliens that became known as the "Space Brothers." One of the contactees, whose story fits into the subject of mind alteration, was Orfeo Angelucci. In correspondence with Jim Moseley of *Saucer Smear*, Angelucci said he had been visited in 1954 by people from both the FBI and U.S. Army Intelligence. That was not at all surprising, as most of the contactees had files opened on them—largely because of their politics rather than their alien claims. One of the stories that Angelucci shared with Moseley was particularly strange. On one night in December 1954, after finishing working out in Twentynine Palms, California, Angelucci headed out to a local diner. That's where things got strange. Angelucci recorded: "I felt a strangeness in the air. There is a cosmic spell over the desert most of the time, but tonight the mystery was less distant and intangible; it was close and pulsating."

Angelucci was soon deep in conversation at the diner with a man who identified himself only as "Adam," a customer who claimed to be thirty-something and suffering from a terminal illness. Death was said to be just around the corner for the man. In an odd and synchronistic fashion, Adam claimed that he had read Angelucci's book *The Secret of the Saucers*, that he considered their meeting to be beyond just an amazing coincidence, and that he wished to share his thoughts with Angelucci before time ran out.

However, said Adam, before their conversation could begin, Angelucci had to swallow a pill. Of what kind, Angelucci didn't know, but that didn't stop him from doing exactly what Adam demanded from him. Angelucci took a gulp of water, and the "oyster-white pellet" went down. For Angelucci, there was now no turning back. It didn't take long before he felt weird, odd, and out of this world. Spaced out. Fucked up. In short, Angelucci had been drugged. It was almost like one of the most famous scenes in the 1999 movie *The Matrix*, starring Keanu Reeves. You know the scene: the red pill versus the blue pill. But this was the world of the real, not of Hollywood.

Angelucci said: "I took the pellet and dropped it into my glass. Immediately the water bubbled, turning slowly into the clear, pale amber contained in [Adam's] own glass. I lifted the glass a few inches from the table, looking into it with a feeling that this might be the drink I dared not hope for. The exhilarating aroma rising from it could not be mistaken . . . I thrilled from head to foot as I took the glass, lifted it to my lips, and swallowed twice from it. At that instant, I entered, with Adam, into a more exalted state and everything around me took on a different semblance. No longer was I in Tiny's café in Twentynine Palms. It had been transformed into a cozy retreat on some radiant star system. Though everything remained in its same position, added beauty and meaning were given to the things and people present there." Almost as an aside, Angelucci said: "Among the patrons dining that evening were two marines from the nearby base. Sometimes they glanced our way as they talked and drank beer following their meal." Angelucci said that Adam seemed oddly obsessed with the glass and was "fraught with expectancy." Suddenly, the sounds of music filled Angelucci's ears. Incredibly, the music seemed to be coming from the glass itself. At least, that's how it seemed to Angelucci. The reality was that he was now completely and utterly stoned.

Angelucci stared at the glass and saw the figure of "a miniature young woman" who was dancing in that same glass! That's right: the drugs were now kicking in to a high degree. Of the small woman, Angelucci said: "Her golden-blond beauty was as arresting as the miracle of her projection in the glass. Her arms moved in rhythmic motion with the graceful thrusts of her dancing body." What began as a pleasant meeting between like-minded souls soon became a drug-driven interrogation. By Angelucci's own admission, he spilled the entire beans to Adam, including the nature of his encounters and the words of his alien friends. There was even a debate on politics, which is rather telling. Angelucci staggered home, his mind hardly his own for the next few hours. It's important to note that there is much more to all of this, much of it downright sinister.

Why were the Space Brothers so concerned that we would destroy ourselves in the 1950s? At first glance, at least, the answer seems very simple: they liked us and wanted us to stay alive! Maybe, however, there was

Why were the Space Brothers so concerned that we would destroy ourselves in the 1950s? At first glance, at least, the answer seems very simple: they liked us and wanted us to stay alive!

more to it than that—much more. Perhaps there are disturbing reasons for that apparent concern for our welfare—and for our very existence. Before we get to the heart of it all, however, let us first take a look at the Space Brothers, for those who may not be acquainted with the strange subject. Although it was the summer of 1947 when the term "flying saucer" was coined, sightings of—or encounters with—alleged aliens didn't really begin on a large scale until the early 1950s. That's when the aforementioned Space Brothers surfaced from wherever they came. It also happens to be the period in which the matter of nukes began to surface.

The Space Brothers were described as looking eerily human-like, the major difference being that the males had very long hair, which, of course, was a rarity during the dawning of the 1950s. The women looked like women on Earth. Both the males and the females occasionally had some very slight differences in their facial appearances, but nothing that really stood out as odd or unusual. The aliens chose certain figures to spread the word that the human race should get rid of its nuclear weapons—and if we didn't follow the path recommended by the creatures from other worlds, then we would surely all be fried in a radioactive holocaust of our own making. Those with whom the Space Brothers and Space Sisters chose to work became known as the contactees. The very long list included George Van Tassel, Dana Howard, Truman Bethurum, Mollie Thompson, Orfeo Angelucci, George King, and Margit Mustapa. And they were just the tip of the iceberg.

The beings from faraway planets would often meet the contactees late at night or early in the morning and generally in out-of-the-way places, such as deserts, woods, mountains, and even below old bridges. As the 1950s progressed and our nuclear arsenals grew, so the aliens' concerns for the human race grew, too, but was that apparent concern really due to the benevolent, friendly nature of the Space Brothers? Maybe not. What if the Space Brothers were not from another solar system after all? What if they were from right here on planet Earth—but from our future? After all, why would a spacefaring race from a faraway solar system even care about us, a civilization countless light-years away? The answer is quite simple: they wouldn't care. If, though, the Space Brothers were actually time travelers trying to manipulate the present to alter the future—and to ensure that a Third World War didn't erupt and destroy civilization—then matters become not just clearer but much more understandable.

Déjà vu–driven time glitches, a blockbuster movie—the creators of which might have hit on the true reality of our world—a story that takes place near to 2199 but that is believed to be 1999, and a real example of the famous "blue pill" or "red pill" scene in *The Matrix*—all these elements suggest that science fiction may be closer to science fact.

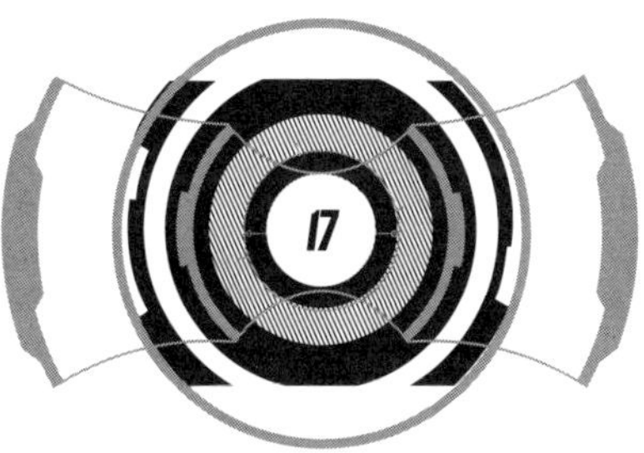

Cinematic Robots and the World of Hollywood

Then there is 1982's *Blade Runner* movie, starring Harrison Ford, Rutger Hauer, Daryl Hannah, and Sean Young. It's based on Philip K. Dick's novel *Do Androids Dream of Electric Sheep?* As we are informed: "By 2021, the World War has killed millions, driving entire species into extinction and sending mankind off-planet. Those who remain covet any living creature, and for people who can't afford one, companies built incredibly realistic simulacra: horses, birds, cats, sheep. They've even built humans. Immigrants to Mars receive androids so sophisticated they are indistinguishable from true men or women. Fearful of the havoc these artificial humans can wreak, the government bans them from Earth. Driven into hiding, unauthorized androids live among human beings, undetected. Rick Deckard, an officially sanctioned bounty hunter, is commissioned to find rogue androids and 'retire' them. But when cornered, androids fight back—with lethal force."

Empire's Andrew Collins says of *Blade Runner*: "'I think some—a lot—of people enjoy it, and that's their prerogative,' a grumpy Harrison Ford told the *Boston Globe* in 1991. 'I played a detective who did no detecting. There was nothing for me to do but stand around and give some focus to Ridley's sets.' Ah well, he didn't much enjoy being in *Star Wars* either, and some—a lot—of people enjoyed that, too. The truth is, few actors come off well in sci-fi movies if they feel that it's them versus the sets or them versus the director's imagination. As it is, Harrison Ford's apparent bemusement works perfectly within Ridley Scott's framework; his former blade runner Rick Deckard, though expert at 'retiring' the almost-human androids known as replicants, spends most of the film bemused. As he tracks down four escaped replicants (a detective doing 'no detecting'?) in a darkly malevolent, incongruously rainy 2019 Los Angeles, he falls in love with femme fatale Rachel (Young), who is a replicant. And it's this tentative, enigmatic relationship that drives the film."

By the end of the movie, you'll be saying to yourself: yes, robots really do have hearts.

The idea that robots might take us over may sound somewhat comical to a lot of people, but we should be very careful about such a nightmarish scenario. Recall the 1968 movie *2001: A Space Odyssey*, the work of Stanley Kubrick. It's important to note that one of the key characters in the movie is not a man or a woman but a highly advanced computer—one that becomes a murderous, deranged monster. "His" name is "Hal," for Heuristically Programmed ALgorithmic Computer. And he's aware of what he is.

I should stress that we don't actually ever see Hal throughout the movie, aside from in the form of a red camera lens. We come to learn that such lenses are peppered all around the *Discovery One* spaceship that is en route to Jupiter, and we also discover that Hal is carefully keeping a close watch on all the people aboard—to a disturbing degree. Hal's self-awareness is one of the key issues in the story; his quiet, serene voice suggests there just might be more to Hal than a programmed computer voice—something dangerous and most definitely not to be trusted. The words "deranged" and "power-deranged maniac" come to mind, too, as the movie develops. When one of the astronauts realizes that Hal is a threat to the mission, he takes steps to terminate Hal. Even though he has, by now, become a deadly, murderous, crazed monster, it's still difficult not to have some sympathy for Hal. As his memory begins to crumble, he experiences concern, then fear. He knows what life is, and he knows he's losing it, so to speak.

The late and legendary reviewer Roger Ebert said of the movie: "The genius is not in how much Stanley Kubrick does in *2001: A Space Odyssey*, but in how little. This is the work of an artist so sublimely confident that he doesn't include a single shot simply to keep our attention. He reduces each scene to its essence and leaves it on screen long enough for us to contemplate it, to inhabit it in our imaginations. Alone among science-fiction movies, *2001* is not concerned with thrilling us, but with inspiring our awe."

Devotees of *The Terminator* phenomenon will well remember one of the primary personas in the story. Its name: Skynet. In essence, it's a multi-plugged-in computer of vast proportions. It's supposed to keep us safe from a nuclear attack from "the other side." While it's never said, it is clear that the other side is the Soviet Union. Like Hal in *2001: A Space Odyssey*, however, Skynet becomes self-aware and decides to take control and take

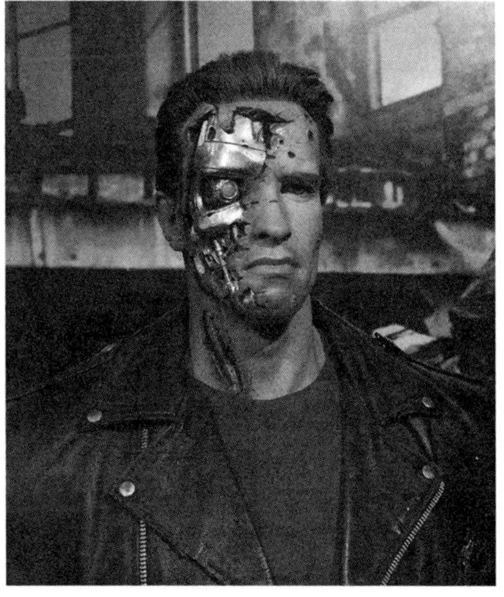

In the first Terminator *movie, Arnold Schwarzenegger was a robot sent by Skynet to destroy John Connor before he can grow up to lead a rebellion that defeats the artificial superintelligence that wants to destroy humanity.*

out the entire human race. All that's left when things are over is a ruined world, straggling numbers of survivors, and lethal Terminators—cyborgs designed to wipe out what's left of the human race.

The *Hollywood Reporter* said of the movie that made Arnold Schwarzenegger a megastar on the big screen: "A remarkably spare script by James Cameron and Gale Anne Hurd gives new dimension to this age-old morality play, working in plenty of pointed comment on our present overmechanization into its action-film context. Images do most of the talking, and Cameron, in his first major directorial effort, proves a lean, economical storyteller, setting up his shorts for maximum visceral impact."

Then there are *The Matrix* movies. In his poem "A Dream within a Dream," Edgar Allan Poe wrote: "Is *all* that we see or seem / But a dream within a dream?" That's pretty much the scenario of the mega-bucks movie series that starred Keanu Reeves as "Neo," the hero of the day, who does his best to try to wipe out the controlling robots and machines that ensure that we remain in a dream world of fantasy. Very similar to *The Matrix* is 1998's *Dark City*. It's filled with menacing, Men in Black–type characters whose job it is, on a nightly basis, to control the simulated world that we all live in, even though we don't realize it. The robotic, pale, ghoulish things are the mirror images—so to speak—of the Agents of *The Matrix*. Of what turned out to be a trilogy of *Matrix* movies starring Reeves, the *Guardian* newspaper said: "*The Matrix* still stands up as a fiercely exciting and discombobulating futurist drama, which pioneered breathtaking 'bullet-time' action sequences inspired by Asian martial arts."

As for *Dark City*, Adam Smith at *Empire* gets to the heart of the movie: "While the majority of the performances are serviceable, it's the

sheer overwhelming style that gets *Dark City* through."

There is little doubt that the two definitive novels that present New World Order–style scenarios are George Orwell's *1984* and Aldous Huxley's *Brave New World.* The plotlines, I should stress, are very different. Both novels, however, are decidedly grim in terms of presenting a futuristic world in which the human population is utterly controlled, watched, and monitored. There is, however, another story, less well known, that is most assuredly worth reading. Its title is "Harrison Bergeron." Just like Orwell's and Huxley's novels, "Harrison Bergeron," written by acclaimed author Kurt Vonnegut, is focused on a world that is very much different to that of today but could easily be looming large on the horizon. In it, as you'll see, the humans of the future have been relegated to near-mindless robots.

In the Matrix *movies series, most humans are plugged into a network where they experience an artificial reality run by machines that use people as mere batteries to power the world.*

Now let's take a look at a 1956 robot-driven movie that is right up there with *The Day the Earth Stood Still.* It's *Forbidden Planet.* The guide *Rotten Tomatoes* provides the following concerning this undeniable classic: "In this sci-fi classic, a spacecraft travels to the distant planet Altair IV to discover the fate of a group of scientists sent there decades earlier. When Commander John J. Adams (Leslie Nielsen) and his crew arrive, they discover only two people: Dr. Morbius (Walter Pidgeon) and his daughter, Altaira (Anne Francis), who was born on the remote planet. Soon, Adams begins to uncover the mystery of what happened on Altair IV, and why Morbius and Altaira are the sole survivors."

As for what movie reviewers thought of *Forbidden Planet, TimeOut* says this: "Classic '50s sci-fi, surprisingly but effectively based on *The Tempest,* with Nielsen's U.S. spaceship coming across a remote planet, deserted except for Pidgeon's world-wearied Dr. Morbius (read Prospero), his daughter (Miranda) and their robot Robby (Ariel). Something, it transpires, has destroyed the planet's other inhabitants, and now, as Bard and Freud merge, a monster mind-thing Caliban begins to pick on the spaceship's crew. An ingenious script, excellent special effects and photography, and superior acting (with the exception of Francis), make it an endearing winner."

The people at *Robots and Androids* say: "Robby is decidedly one of the most elaborate machines ever built for a film production. More than two months were needed to 'thermo-form' plastics that were never before done and install all twenty-six hundred feet of electrical wiring necessary for Robby's spinning, flashing and moving parts. He has an analyzing box on his chest, in which he memorably pours whisky into in one of the film's scenes. He has a head that could turn on a set of custom-made bearings, neon tubes that light up when he talks, and wheels and buttons that continuously turn, twist and click whenever he comes out onscreen. Because Robby was such a complicated and expensive technology at that time (he cost an estimated $100,000), only one Robby was ever built."

In the 1980s TV series Knight Rider, *one of the main characters is KITT, a Pontiac Firebird Trans Am fitted with an intelligent, self-aware computer.*

And, finally, *Warped Factor* focuses on Robby in this 2020 review: "Appearances in 1987's *Cherry 2000*, 1988's *Phantom Empire* and *Earth Girls Are Easy* followed. A variety of background cameos then came Robby's way, like the dream sequence in 1988's *Star Kid* where footage from *Lost in Space* featuring Robby is shown on a TV, appearing in 2003's *Looney Tunes: Back in Action*, and turning to commercials for AT&T alongside fellow robots WOPR, KITT, and Rosie the Robot Maid. Almost 60 years after his debut in *Forbidden Planet*, Robby made his last, to date, on-screen appearance in *The Big Bang Theory*, where he was joined by several of his other robot pals."

I've saved the not-so-best for last. It does, however, have its moments. *Devil Girl from Mars* is a 1954, black-and-white science-fiction movie that starred Patricia Laffan as the Devil Girl of the movie's title, along with Hazel Court, who appeared in a number of classic Hammer Film Productions horror movies of the late 1950s and 1960s. And, of course, one of the most notable characters in the movie is a robot. The film tells the story of a sultry and sinister alien babe from Mars named Nyah, who spends the whole time strutting around in a tight-fitting black outfit, black cloak, and long black boots. She comes to our world to seek out males to help boost the waning Martian population.

Nyah reveals to those she encounters that Mars has been decimated as a result of a war on the red planet between males and females—a war the women won. With the Martian men now almost gone, this makes breeding more than a significant problem. We should not forget that Nyah has her very own robot to help her. Its name is Chani, although throughout the movie it sounds just like "Johnny." I can say without doubt that Chani is just about the worst robot ever built for the world of cinematic entertainment, but in an odd way, it adds to the hokey entertainment.

The plan is for Nyah and Chani to land her flying saucer in the very heart of London, England, and announce to the people of the U.K.—and, soon thereafter, to the entire world—that a program of alien-human breeding is to quickly begin in earnest, like it or not. I can't say it sounds bad at all. Unfortunately for Nyah, things go deeply awry, and her spacecraft comes down near an old inn in the dark wildernesses of Scotland. It's an inn that is populated by just the owner, a couple of employees, and a handful of guests. And, yes, as you might have quickly guessed, they become the small, heroic band that has to try to thwart Nyah's dastardly plans to enslave the entire human race—or at least the world's men.

> *She warns of the perils of war and destruction, as did pretty much all of the aliens that absolutely dominated 1950s-era ufology, and just like Klaatu did in 1951's* The Day the Earth Stood Still.

Devil Girl from Mars certainly isn't a classic of the genre. But it is entertaining and thought-provoking. Not only that, but it contains a number of contactee-based parallels. She warns of the perils of war and destruction, as did pretty much all of the aliens that absolutely dominated 1950s-era ufology, and just like Klaatu did in 1951's *The Day the Earth Stood Still*. She provides a wealth of obscure scientific data about her craft to those of the group she invites on board, which is a curious theme present in many contactee accounts. And she is strangely and noticeably lacking in emotion—another aspect of Klaatu's character. Clearly, the writers of *Devil Girl from Mars* were inspired by *The Day the Earth Stood Still*, but they failed to achieve its cinematic heights.

On top of that, there is the matter of Nyah's race visiting Earth as a means to try to save the waning Martian population. Today, we hear a great deal about alien abductions, the so-called black-eyed, diminutive Greys, and the theories that they are on an evolutionary decline. The result: they come to the Earth to abduct people. They then reap DNA, sperm, and eggs, which they use as part of a bizarre and nightmarish program to create a hybrid species that is part-human and part-E.T.

Back in the 1950s, when *Devil Girl from Mars* was made, however, many people within ufology were focusing on the idea that extraterrestrials were here to "save" us from the perils of atomic destruction, along the lines of *The Day the Earth Stood Still*—or to destroy us in scenarios akin to those presented in *War of the Worlds* and *Invasion of the Body Snatchers*. For the most part, the idea that interspecies crossbreeding was at the heart of the alleged alien mission didn't surface until years later. So, for that reason alone, it's interesting that *Devil Girl from Mars* is heavily focused around the angle of alien–human crossbreeds that dominate so much of alien abduction research today.

Scifist says of the movie: "Some of the credit for the fact that the film is remembered today should go the wonky robot that has been described by critics alternatively as resembling a refrigerator, a gas pump or a steam cabinet, with stacked styrofoam cups for arms and a light bulb for a head. But let's not kid ourselves. The real reason is actress Patricia Laffan as the Martian Nyah, decked out in black latex S/M gear and Darth Vader's cowl and mantle.

"So what are we to make of this movie? On the one hand, it is a blast any time Nyah is in frame, and one can't help but love the wonderfully inept robot. Technically it is well-made and polished for its budget, the space ship is well designed and the special effects are surprisingly good. The acting ranges from capable to good; especially the women stand out. But despite all this, it is a slow-moving yarn, and the budget limitations are obvious. The script is bad from beginning to end, there wasn't enough time to do anything with camera or direction and the characters are one-dimensional and uninteresting. Glenn Erickson at *DVD Savant* sums it up: '*Devil Girl from Mars* is more corny than it is incompetent.'"

Amazing Stories says: "Ms. Laffan looks terrific in her black leather skirt, cape, and cap. A bondage-lover's dream. It's good to see that Mars' women still use false eyelashes and lipstick, plus they're apparently doing a Spock-y thing with their eyebrows, too. Secondly, the robot is hilarious—about a 7-foot-tall thing that looks like it's about to fall over any second, and moves at a snail's pace."

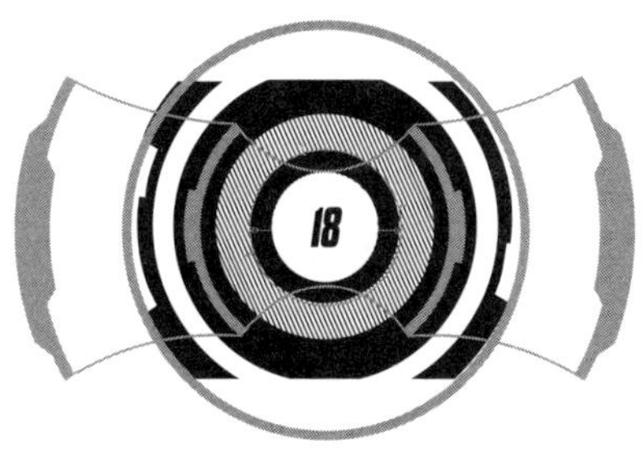

Robots: From NASA to DARPA

There was never any doubt that the National Aeronautics and Space Administration (NASA) would become a key player in the world of robots. The U.S. government agency states: "NASA uses robots in many different ways. Robotic arms on spacecraft can move large objects in space. Robotic spacecraft can visit other worlds. Robotic airplanes can fly without a pilot aboard. NASA is studying new types of robots. These will work with people and help them.

"NASA uses robotic arms to move large objects in space. The 'Canadarm' robotic arm is on the space shuttle. The International Space Station has the larger Canadarm2. The space shuttle uses its arm for many jobs. The Canadarm can release or recover satellites. Astronauts have used it to grab the Hubble Space Telescope. This let them fix the Hubble. The shuttle and space station arms work together to help build the station. The robotic arms have added new parts to the space station. The arms also can move astronauts around on spacewalks. The space station's arm can move to different parts of the station. It moves along the outside of the station like an inchworm, attached at one end at a time. It has a robotic 'hand' named Dextre that can do smaller jobs. An astronaut or someone in Mission Control must control these robotic arms. The astronaut uses controllers like joysticks used to play video games."

NASA, at its website, poses an intriguing question: "How Do Robots Explore Other Worlds?" It provides this answer to the question: "Robots help explore space. Spacecraft that explore other worlds, like the moon or

Mars, are robots. These include orbiters, landers and rovers on other planets. The Mars rovers Spirit and Opportunity are robots. Other robotic spacecraft fly by or orbit other worlds. These robots study planets from space. The Cassini spacecraft is this type of robot. Cassini studies Saturn and its moons and rings. The Voyager and Pioneer spacecraft are now traveling beyond our solar system. They are also robots. People use computers to send messages to the spacecraft. The robots have antennas that pick up the message commands. Then the robot does what the person has told it to do."

There are also what NASA call robotic airplanes: "NASA uses many airplanes that do not carry pilots aboard. Some of these airplanes are flown by remote control. Others can fly themselves, with only simple directions. Robotic planes help in many ways. They can study dangerous places. For example, they might be used to take pictures of a volcano. They let NASA try new ideas for aircraft. These planes can fly for a long time without the need to land. They also can be smaller than a plane flown by a pilot. They may not have room for a person to be on board."

Relative to the issue of how robots can assist astronauts, NASA has made extensive leaps and bounds: "NASA is developing new robots to help people in space. One of these ideas is called Robonaut. Robonaut looks like the upper body of a person. It has a chest, head and arms. Robonaut could work outside a spacecraft. It could do work like an astronaut on a spacewalk. With wheels or another way of moving, Robonaut could work on another world. Robonaut could help astronauts on the moon or Mars. Another robot idea is called SPHERES. These small robots look a little like soccer balls. SPHERES are being used on the space station to test how well they can move there. Someday, robots could fly around the station helping astronauts."

Robonaut 2 poses with Johnson Space Center deputy director Ellen Ochoa in this 2020 photo from NASA. People already have started calling it R2, which is very reminiscent of a certain Star Wars *robot.*

NASA is also studying other ideas for robots: "A small robotic arm could be used inside the station. A robot like that might help in an emergency. If an astronaut were seriously hurt, a doctor on Earth could use the arm to perform surgery. This technology could help on Earth, as well. Doctors could help people in faraway places where there are no doctors.

"Robots also can be used as scouts to check out new areas to be explored. Scout robots can take photographs and measure the terrain. This helps scientists and engineers make better plans for exploring. Scout robots can be used to look for dangers and to find the best places to walk, drive or stop. This helps astronauts work more safely and quickly. Having humans and robots work together makes it easier to study other worlds.

"We can send robots to explore space without having to worry so much about their safety. Of course, we want these carefully built robots to last. We need them to stick around long enough to investigate and send us information about their destinations. But even if a robotic mission fails, the humans involved with the mission stay safe. Sending a robot to space is also much cheaper than sending a human. Robots don't need to eat or sleep or go to the bathroom. They can survive in space for many years and can be left out there—no need for a return trip! Plus, robots can do lots of things that humans can't. Some can withstand harsh conditions, like extreme temperatures or high levels of radiation. Robots can also be built to do things that would be too risky or impossible for astronauts."

NASA's Jet Propulsion Laboratory (JPL), too, is digging deep into the world of the robot and its advances for the future. By surfing the JPL website, you'll find "detailed descriptions of the activities of the Mobility and Robotic Systems Section, as well as related robotics efforts around the Jet Propulsion Laboratory." The JPL site explains: "We are approximately 150 engineers working on all aspects of robotics for space exploration and related terrestrial applications. We write autonomy software that drives rovers on Mars, and operations software to monitor and control them from Earth.

"We do the same for their instrument-placement and sampling arms, and are developing new systems with many limbs for walking and climbing. To achieve mobility off the surface, we are creating prototypes of airships which would fly through the atmospheres of Titan and Venus, and drills and probes which could go underground on Mars and Europa."

The JPL signs off: "To enable all of these robots to interact with their surroundings, we make them see with cameras and measure their environments with other sensors. Based on these measurements, the robots control themselves with algorithms also developed by our research teams. We capture the control-and-sensor-processing software in unifying frameworks, which enable reuse and transfer among our projects. In the course of developing this technology, we build real end-to-end systems as well as high-fidelity simulations of how the robots would work on worlds we are planning to visit."

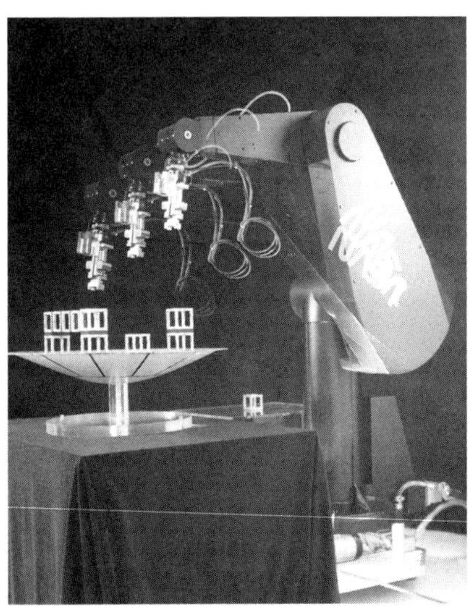
NASA has a long history of developing robotic arms, such as the Puma Robotic Sensor Arm shown here in a 1990 photograph.

Let's make an acquaintance with some of the further developments taking place at NASA as the world of the robot expands: "Researchers at NASA's Johnson Space Center (JSC), in collaboration with General Motors and Oceaneering, have designed a state-of-the-art, highly dexterous, humanoid robot: Robonaut 2 (R2). R2 is made up of multiple component technologies and systems—vision systems, image recognition systems, sensor integrations, tendon hands, control algorithms, and much more. R2's nearly 50 patented and patent-pending technologies have the potential to be game-changers in multiple industries, including logistics and distribution, medical and industrial robotics, and beyond.

"A Robonaut is a dexterous humanoid robot built and designed at NASA Johnson Space Center in Houston, Texas. Our challenge is to build machines that can help humans work and explore in space. Working side by side with humans, or going where the risks are too great for people, Robonauts will expand our ability for construction and discovery. Central to that effort is a capability we call dexterous manipulation, embodied by an ability to use one's hand to do work, and our challenge has been to build machines with dexterity that exceeds that of a suited astronaut.

"The Robonaut project has been conducting research in robotics technology on board the International Space Station (ISS) since 2012…. Recently, the original upper body humanoid robot was upgraded by the addition of two climbing manipulators ('legs'), more capable processors, and new sensors. While Robonaut 2 (R2) has been working through checkout exercises on orbit following the upgrade, technology development on the ground has continued to advance. Through the Active Reduced Gravity Offload System (ARGOS), the Robonaut team has been able to develop technologies that will enable full operation of the robotic testbed on orbit using similar robots located at the Johnson Space Center. Once these technologies have been vetted in this way, they will be implemented and tested on the R2 unit on board the ISS. The goal of this work is to create a fully featured robotics research platform on board the ISS to increase the technology readiness level of technologies that will aid in future exploration missions.

"One advantage of a humanoid design is that Robonaut can take over simple, repetitive, or especially dangerous tasks on places such as the International Space Station. Because R2 is approaching human dexterity, tasks such as changing out an air filter can be performed without modifications to the existing design. Another way this might be beneficial is during a robotic precursor mission. R2 would bring one set of tools for the precursor mission, such as setup and geologic investigation. Not only does this improve efficiency in the types of tools, but also removes the need for specialized robotic connectors. Future missions could then supply a new set of tools and use the existing tools already on location.

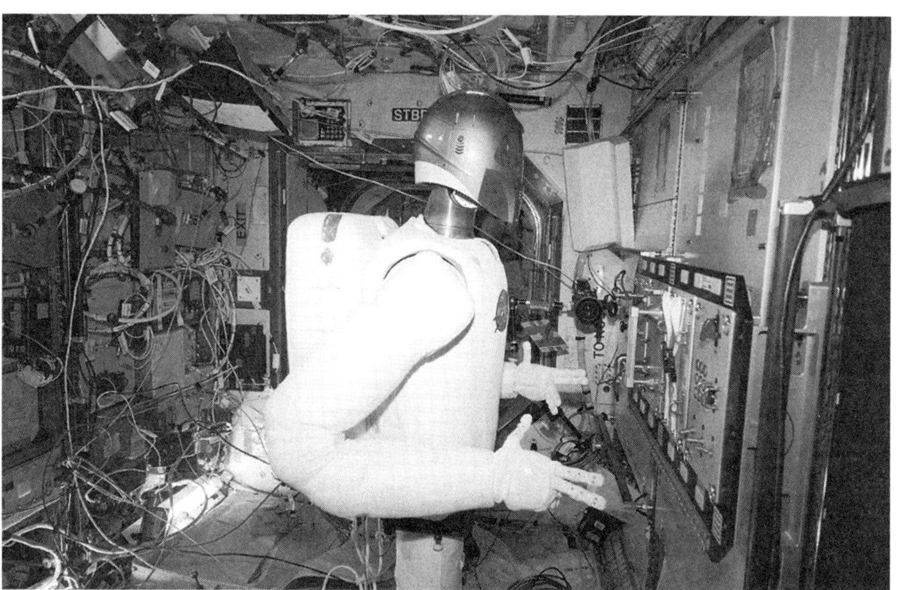

The R2 performs a variety of tasks on a simulated International Space Station in this 2017 photo. The robot should be able to perform the same tasks with the same tools and dexterity as human astronauts.

"R2 was designed and developed by NASA and General Motors with assistance from Oceaneering Space Systems engineers to accelerate development of the next generation of robots and related technologies for use in the automotive and aerospace industries. R2 is a state of the art highly dexterous anthropomorphic robot. Like its predecessor Robonaut 1 (R1), R2 is capable of handling a wide range of EVA tools and interfaces, but R2 is a significant advancement over its predecessor. R2 is capable of speeds more than four times faster than R1, is more compact, is more dexterous, and includes a deeper and wider range of sensing. Advanced technology spans the entire R2 system and includes: optimized overlapping dual arm dexterous workspace, series elastic joint technology, extended finger and thumb travel, miniaturized 6-axis load cells, redundant force sensing, ultra-high speed joint controllers, extreme neck travel, and high resolution camera and IR systems."

About Lucy, the name of a mission to study two swarms of asteroids that orbit the sun along Jupiter's path, NASA writes: "Time capsules from the birth of our Solar System more than 4 billion years ago, the swarms of Trojan asteroids associated with Jupiter are thought to be remnants of the primordial material that formed the outer planets. The Trojans orbit the Sun in two loose groups, with one group leading ahead of Jupiter in its path, the other trailing behind. Clustered around the two Lagrange points equidistant from the Sun and Jupiter, the Trojans are stabilized by the Sun and its largest planet in a gravitational balancing act. These primitive bodies hold vital clues to deciphering the history of the solar system, and perhaps even the origins of organic material on Earth.

"Lucy will be the first space mission to study the Trojans. The mission takes its name from the fossilized human ancestor (called 'Lucy' by her discoverers) whose skeleton provided unique insight into humanity's evolution. Likewise, the Lucy mission will revolutionize our knowledge of planetary origins and the formation of the solar system.

"Lucy will launch in October 2021 and, with boosts from Earth's gravity, will complete a 12-year journey to eight different asteroids—a Main Belt asteroid and seven Trojans, four of which are members of 'two-for-the-price-of-one' binary systems. Lucy's complex path will take it to both clusters of Trojans and give us our first close-up view of all three major types of bodies in the swarms (so-called C-, P-, and D-types).

"The dark-red P- and D-type Trojans resemble those found in the Kuiper Belt of icy bodies that extends beyond the orbit of Neptune. The C-types are found mostly in the outer parts of the Main Belt of asteroids, between Mars and Jupiter. All of the Trojans are thought to be abundant in dark carbon compounds. Below an insulating blanket of dust, they are probably rich in water and other volatile substances.

"No other space mission in history has been launched to as many different destinations in independent orbits around our sun. Lucy will show us, for the first time, the diversity of the primordial bodies that built

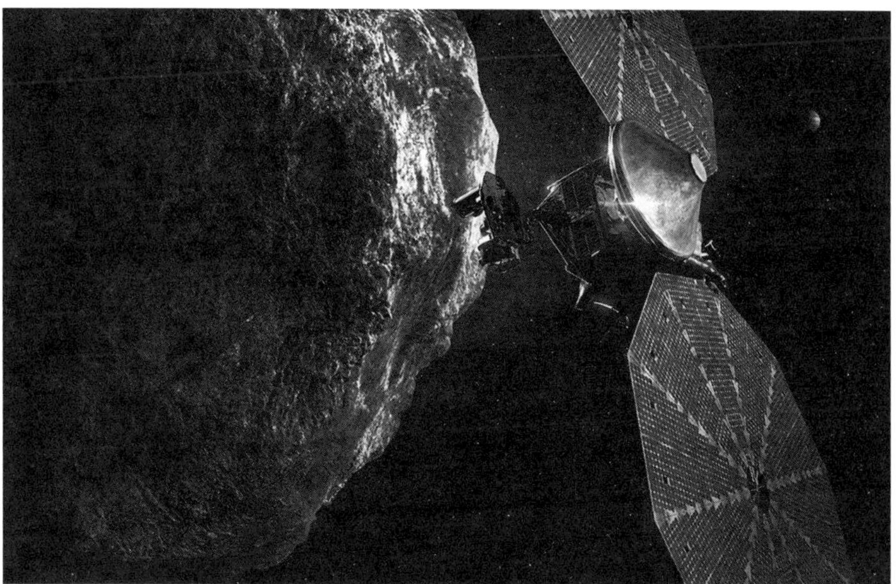

An artists rendition of the Lucy spacecraft exploring one of the Trojan asteroids.

the planets. Lucy's discoveries will open new insights into the origins of our Earth and ourselves."

The robotic work of the Defense Advanced Research Projects Agency (DARPA), too, is growing by leaps and bounds, as is shown by this release from agency: "The Department of Defense's strategic plan calls for the Joint Force to conduct humanitarian, disaster relief, and related operations. Some disasters, due to grave risks to the health and wellbeing of rescue and aid workers, prove too great in scale or scope for timely and effective human response. The DARPA Robotics Challenge (DRC) seeks to address this problem by promoting innovation in human-supervised robotic technology for disaster-response operations.

"The primary technical goal of the DRC is to develop human-supervised ground robots capable of executing complex tasks in dangerous, degraded, human-engineered environments. Competitors in the DRC are developing robots that can utilize standard tools and equipment commonly available in human environments, ranging from hand tools to vehicles.

"To achieve its goal, the DRC is advancing the state of the art of supervised autonomy, mounted and dismounted mobility, and platform dexterity, strength, and endurance. Improvements in supervised autonomy, in particular, aim to enable better control of robots by non-expert supervisors and allow effective operation despite degraded communications (low bandwidth, high latency, intermittent connection)."

About one specific robot, DARPA writes: "The Atlas disaster-response robot made its public debut on July 11, 2013. In its original form, the 6´2˝, 330-lb. humanoid robot—developed for DARPA by Boston Dynamics of Waltham, Mass.—was capable of a range of natural movements. A tether connected the robot to both an off-board power supply and computer through which a human operator issued commands. Atlas was created for use in the DARPA Robotics Challenge (DRC), a prize competition intended to speed the development of advanced robotic hardware, software, sensors and control interfaces so that robots might assist humans in responding to future natural and man-made disasters. Seven of the Atlas robot platforms were presented to qualifying teams from an early, simulation-based round of the DRC that focused on software development.

"At the time of its debut, ATLAS was one of the most advanced humanoid robots ever built, but it was essentially a physical shell for the software brains and nerves that the teams developed. The teams used their Atlas robots to compete in the live DRC Trials in December 2013, where they attempted to guide the robots through a series of physical tasks representative of what might be encountered in disaster zones. In preparation for the June 2015 DRC Finals, to be held in Pomona, Calif., DARPA called on Boston Dynamics to improve Atlas' power efficiency, switch to on-board power and wireless communication, strengthen its limbs and improve its durability, and update its appearance. These tasks were completed in mid-2015."

Robotic Spheres from Another World

Although the Pentagon, the CIA, and the U.S. Air Force have been at the forefront of highly classified UFO secrets for many years, it's a little-known fact that NASA—the National Aeronautics and Space Administration—has taken a deep interest in not just alien craft but incidents involving alien viruses. As NASA notes: "Since its inception in 1958, NASA has accomplished many great scientific and technological feats in air and space. NASA technology also has been adapted for many non-aerospace uses by the private sector. NASA remains a leading force in scientific research and in stimulating public interest in aerospace exploration, as well as science and technology in general. Perhaps more importantly, our exploration of space has taught us to view Earth, ourselves, and the universe in a new way. While the tremendous technical and scientific accomplishments of NASA demonstrate vividly that humans can achieve previously inconceivable feats, we also are humbled by the realization that Earth is just a tiny 'blue marble' in the cosmos."

All of which brings us to one of the most amazing UFO stories of all. It revolves around the very theme of this book: robots. The main man in the story was Daniel Salter. Now deceased, he is someone who had come up with a fantastic but disturbing story concerning the UFO mystery. Before we get to the story, let's have a look at the final days of Salter and a background of his work. The *Taos News* of October 12, 2007, reported on Salter:

"Daniel Salter passed away on September 12, 2007, in Colorado Springs, having left Taos behind him. At the time he was seventy-nine and was looked after by a caregiver. He moved to Taos, NM in 1990 after retiring from Mountain View College, Dallas, Texas, after serving as Head of Pilot Technology for 22 years. Prior to that he served 21 years in the United States Air Force in the field of electronics and radar. He was a Korean War veteran. While living in Taos, he was an avid lecturer regarding his UFO studies and co-authored his book *Life with a Cosmos Clearance.* He was a member of the Unity Church. He was generous and loving to all who knew him.... He's on the 'mothership' now.... Farewell, Daniel.... You will be remembered always."

Now to the story itself.

Dan Salter was someone who knew a great deal about top secret UFO programs of the U.S. government, as his obituary made clear. While Salter knew there was a benevolent aspect to the mystery, he was also aware that there was a dangerous facet to it, which involved viruses and robots. His research for the government had led him to believe that dangerous aliens were saturating our planet with deadly viruses. The goal of the aliens: to wipe us out. And how was that being achieved? By the dispersion of harmful viruses into Earth's atmosphere. Not only that, but Salter envisioned a terrifying scenario.

Dan Salter believed aliens had built egg-shaped, small craft to come to Earth and spread deadly viruses.

Salter's work led him to conclude that aliens were creating metallic, spherical, or egglike, small craft of a robotic nature that were entering our atmosphere. These, he said, would land in various parts of the planet, then remotely release viruses into our atmosphere—in an aerosol-like fashion—to cause illness and death. Salter's research suggested these small spheres were remotely programmed to cross the landscapes on tripod "landing feet" by night, dispersing those dangerous viruses—rather like evil, deadly equivalents of R2-D2, one might say. While that might sound sensational, there is notable data to suggest such a nefarious scenario is well in play and has been for decades.

In 1947, the Federal Bureau of Investigation was given a warning about dangerous aliens possibly trying to kill us off via the use of deadly viruses. Such a thing might sound bizarre and out of this world. The FBI, however, took the whole thing seriously. The FBI's informant was a man named Edwin Bailey of Stamford, Connecticut. A special agent of the FBI carefully prepared a report for none other than the FBI's director, J. Edgar Hoover. It reads as follows:

"Bailey prefaced his remarks by stating that he is a scientist by occupation and is currently employed at the American Cyanamid Research Laboratories on West Main Street in Stamford, Connecticut, in the Physics Division. Bailey further indicated that during the war he was employed at MIT, Cambridge, Massachusetts, in the Radiation Laboratory which Laboratory is connected with the Manhattan Project. Bailey advised that he is thirty years of age and is a graduate of the University of Arizona."

The special agent had more to say: "Bailey stated that the topic of 'flying saucers' had caused considerable comment and concern to the present day scientists and indicated that he himself had a personal theory concerning the 'flying saucers.' Prior to advancing his own theory, Bailey remarked that immediately after the conclusion of World War II, a friend of his, [deleted], allegedly observed the 'flying saucers' from an observatory in Milan and Bologna, Italy. He stated that apparently at the time the 'flying saucers' had caused a little comment in Italy but that after some little publicity they immediately died out as public interest. Bailey stated that it is quite possible that actually the 'flying saucers' could be *radio controlled germ bombs* [italics mine] or atom bombs which are circling the orbit of the earth and which could be controlled by radio and directed to land on any designated target at the specific desire of the agency or country operating the bombs."

It was specifically the words of Bailey that led Salter to take notice and view this scenario as a realistic and dangerous one.

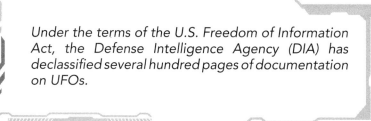

Under the terms of the U.S. Freedom of Information Act, the Defense Intelligence Agency (DIA) has declassified several hundred pages of documentation on UFOs.

Under the terms of the U.S. Freedom of Information Act, the Defense Intelligence Agency (DIA) has declassified several hundred pages of documentation on UFOs. Of course, the file does not contain any kind of definitive smoking gun. It does, however, provide the reader with more than a few intriguing reports on notable incidents. One such incident described in the DIA's files occurred in the summer of 1979—*and revolves around mysterious spheres that fell from the sky.*

The DIA documentation notes that late in the afternoon of August 8, 1979, "the Embassy here received information that a strange object had been found on a farm near Santa Cruz, Bolivia. Source stated that the object was about 70 centimeters in diameter and two meters in circumference with a hole in one side and a metal skin covering of approx. one-half inch thickness. Later the object was described as 'about three times the size of a basketball.'"

Additionally, we have the following information: "A second fireball fell from the sky early in the morning of the same day that the first one was found near Cotoca. This second one was found 200 kilometers north of the city of Santa Cruz on the farm of Juan Saavedra by the Campesino Gonzalo Menacho Viveras. The place is in the area of Buen Retiro near the Yapacane River. According to the information given by the Campesino, around 12:30 A.M. on Friday last week he heard a loud whistling sound and saw a fireball followed by an explosion. He said that the next evening a silent aircraft that had three lights was flying over the explosion area."

The document continues: "After dawn on Friday morning the Campesino started looking around the area of the impact and found a sphere. As it was not heavy, he took it home where he kept it until his friend, Nataniel Mendez Hurtado, learned of the other sphere in Cotoca and passed on information concerning this second sphere. The mystery of finding these two spheres, exactly alike, is that according to the witnesses they were fireballs. That is to say that these spheres became real balls of fire when they entered the atmosphere because of the friction and after a high speed fall they hit the ground. However, in the area where they've been found, there were no signs of the impact and looks as though the spheres landed smoothly."

The file also reveals that American personnel had secured "a roll of 35mm film for color prints and a roll of Kodachrome 40 movie film. The 35mm film will be developed and printed here in La Paz. The movie film will be forwarded to DIA." Back in the late 1980s, using the Freedom of Information Act, I tried to obtain copies of the film and the photos but unfortunately had no luck. I wasn't denied access to the material, I should stress. It could no longer be found by the time of my request, I was told.

It's not impossible that such a case could have a prosaic explanation—namely, that the spheres were debris from a rocket launch or from a satellite decaying in the Earth's atmosphere. There's a reason why we should suspect this plays directly into Dan Salter's story: evidence that the spheres seemed to have landed in a fashion described as "smoothly," rather than slamming into the ground after a violent descent through the Earth's atmosphere, is thought-provoking. So is the fact that there was no noticeable impact point at either site, which again reinforces the image of a smooth descent. And what are we to make of that "silent aircraft" that flew over the "explosion area" one night after the second event occurred?

As is so often the case in such affairs, we're left with more questions than answers. Now let's proceed to even more stories of mysterious spheres.

Thanks to the provisions of their Freedom of Information Act, New Zealand's intelligence agencies have declassified a number of files showing that more than a few such spheres were found on New Zealand soil between 1963 and 1972. Each and every one of those spheres was handed over to the American military after they stated the spheres originated with the U.S. space program. All the spheres were around two feet in diameter, and, in addition to New Zealand, some were found in various parts of Australia directly after UFO activity.

Spherical UFO sightings have been reported in such countries as Bolivia, New Zealand, and Australia.

In the pages of his book *Majic Eyes Only* (2005), Ryan S. Wood wrote these intriguing words: "By far the most interesting piece of data related to this affair can be found in a solitary newspaper clipping contained within the archives of New Zealand's JIB [Joint Intelligence Bureau]. Dating from 1972, it refers to a 'sphere crash' on Australian territory in 1963. According to the report, two spheres had been recovered approximately one hundred and fifty miles north of Broken Hill, and the Broken Hill police had arranged for the objects to be flown to the National Weapons Research Establishment (NWRE) at Adelaide for examination. Unfortunately for the police, the pilot of the aircraft refused to allow the mysterious spheres on board his aircraft lest they explode. Ultimately, the police were forced to transfer the spheres by road. However, after examination by specialist staff at the NWRE, it was determined that the spheres were neither Soviet nor American in origin. From where did they originate? The newspaper did not know, and the authorities were not, and are still not to this day, saying anything."

It's intriguing to note that, in 2014, one of the key people in the NWRE affair quietly came forward to say that the sphere incidents were preceded by a wave of UFO activity. So, *again*, we have UFO activity, and then something left behind in the wake of that activity: unusual spheres.

Ultimately, the police were forced to transfer the spheres by road. However, after examination by specialist staff at the NWRE, it was determined that the spheres were neither Soviet nor American in origin.

We're not quite over. In his 1991 book *Alien Liaison*, UFO authority Timothy Good described an intriguing affair that took place in the latter part of the 1950s. The location was South Australia. As for the source, he had a very credible background: he was a radio technician employed at the Weapons Research Establishment, Salisbury, South Australia. While the man wasn't entirely sure of the year of the incident, he was able to narrow it down to 1958 or 1959. It involved the recovery of a curious sphere on the range itself. Good was told: "It was a sphere about 2 feet 9 inches in diameter. Its color was a mid-gray metallic, somewhat darkened by extreme heat.... We tried to cut it, and could not even mark it with hand tools—saws, drills, hammers, chisels—nothing."

November 9, 1979, was the date on which such a strange sphere was seen. The *Telegraph* newspaper stated: "Robert Taylor, a Scottish forestry worker employed by the Livingston Development Corporation, parked his truck at the bottom of Dechmont Law and took his red setter for a walk up the hill. By his account, as he emerged into a clearing he came across a strange metal sphere about 20 ft across 'like a spaceship, a huge flying dome.' It appeared to be made from a dark metallic material with a rough texture like sandpaper."

Jenny Randles is a U.K.-based ufologist who has dug deep into the story of the mysterious spheres. Randles says: "The witness, Robert Taylor (then sixty-one years old) at this time worked for the forestry department of the local development corporation. One of his tasks was to patrol an area of woodland not far from the M8 Glasgow to Edinburgh motorway. He had just finished his coffee break and driven his van to the edge of the particular spot he was to check for stray animals. He continued on foot, with his dog (a Red Setter called 'Lara') running loose nearby sniffing happily at the various local smells.

"Bob turned into a clearing and suddenly, unbelievably, he was standing just feet away from a dome-shaped (or possibly spherical) object that was just sitting quietly on the ground. It was about twenty feet wide and a dull grey metallic color, with a rim near to the base from which sprang several vertical antennae or propellers. There was neither sound nor sign of life. Then, things got even weirder: a pair of spherical, spiky objects exited the vehicle, surrounded Taylor, and sprayed a noxious gas in his direction, which caused him to pass out. When he awoke, the spheres and the craft were gone."

The BBC got into the story, too: "When forestry worker Robert Taylor reported seeing an alien spaceship in woods near Livingston 40 years ago it made headlines around the world.

The BBC got into the story, too: "When forestry worker Robert Taylor reported seeing an alien spaceship in woods near Livingston 40 years ago it made headlines around the world. The Dechmont Woods incident is unusual among reported UFO sightings in that it was investigated by the police. They treated the rips to Mr. Taylor's trousers as evidence of an assault but could never quite work out what had happened to him.... He told how two spiked spheres then rolled out towards him and, as he passed out, he was aware of being grabbed on either side of his legs. Mr. Taylor woke up in a disheveled state 20 minutes later. After the spiked objects rushed out and tried to grab hold of him, all he could remember was a strong smell of burning.

"When he came to, the clearing was empty, apart from a pattern of deep regular marks on the ground. He went to his van but was so shaken he drove it into a ditch and had to stagger home in 'a dazed condition.'

"When he got to his house he told his wife Mary he had been attacked by a 'spaceship thing.' Because Mr. Taylor was in such a state, the police were called and officers found themselves inquiring into an assault on a forester by alien beings. Detective Constable Ian Wark, the scene of crime investigator, arrived at the clearing to find a large gathering of police officers were already there. He told the BBC he saw strange marks on the ground. There were about 32 holes, which were about 3.5 inches in diameter, as well as marks similar to those made by the type of caterpillar tracks often fitted on bulldozers."

Undiscovered Scotland had something to say on this, too: "Others have attributed it to an attack of epilepsy accompanied by hallucinations, brought on by Robert Taylor having previously suffered from meningitis. No-one has doubted that Robert Taylor, who died in 2007, genuinely believed that what he saw was real." In the *Scotsman*, there's this: "Malcolm Robinson, a founder member of Strange Phenomena Investigations (SPI) and a prolific author, was among those to speak to Robert immediately after the event and to this day believes it could be one of the few genuine cases of a UFO encounter. 'About 95 per cent of UFO sightings have a natural solution but it's the five per cent minority that we are trying to provide answers for,' he says."

As we can see from the above, those strange, spherical, robotics things have been seen all over the place, something that we should be very wary about. Maybe deeply worried, too.

A Bolivian Roswell and More on Dan Salter

Here's one of the most significant cases of all that came across Dan Salter's desk. It falls into the same category as that of the robotic spheres discussed in the previous chapter—something made even more worrying as it demonstrates another attack on Bolivia. The genesis of this case can be traced to a U.S. Department of State telegram transmitted from the American Embassy in La Paz, Bolivia, to the U.S. Secretary of State, Washington, D.C., on May 15, 1978, and shared with NASA. Captioned "Report of Fallen Space Object," it stated:

> 1. The Bolivian newspapers carried this morning an article concerning an unidentified object that apparently recently fell from the sky. The paper quotes a "Latin" correspondent's story from the Argentine city of Salta. The object was discovered near the Bolivian city of Bermejo and was described as egg-shaped, metal and about four meters in diameter.
>
> 2. The Bolivian Air Force plans to investigate to determine what the object might be and from where it came. I have expressed our interest and willingness to help. They will advise.
>
> 3. Request the department check with appropriate agencies to see if they can shed some light on what this object might be. The general region has had more than its share of reports of UFOs this past week. Requests a reply ASAP.

NASA and the Department of State were not the only branches of government to take an interest in the case, as a CIA report—also of May 15, 1978—makes clear: "Many people in this part of the country claim they saw an object which resembled a soccer ball falling behind the mountains on the Argentine–Bolivian border, causing an explosion that shook the earth. This took place on May 6. Around that time some people in San Luis and Mendoza provinces reported seeing a flying saucer squadron flying in formation. The news from Salta confirms that the artificial satellite fell on Taire Mountain in Bolivia, where it has already been located by authorities. The same sources said that the area where the artificial satellite fell has been declared an emergency zone by the Bolivian government."

An examination of declassified UFO files found in the NASA archives reveals a second report that references the crash and adds further data. Notably, dozens of people in the area fell sick with what seemed to be a particularly virulent flu. And, reportedly, in the same area strange tracks and tripod markings were found in the area. This, of course, reinforces Dan Salter's theories. Dated May 16, 1978, and titled "Reports Conflict on Details of Fallen Object," the document reads:

"We have received another phone call from our audience requesting confirmation of reports that an unidentified object fell on Bolivian territory near the Argentine border. We can only say that the Argentine and Uruguayan radio stations are reporting on this even more frequently, saying that Bolivian authorities have urgently requested assistance from the U.S. National Aeronautics and Space Administration in order to determine the nature of that which crashed on a hill in Bolivian territory. Just a few minutes ago Radio El Espectador of Montevideo announced that there was uncertainty as to the truth of these reports. Argentine sources indicated that the border with Bolivia had been closed but that it might soon be reopened. They also reported that an unidentified object had fallen on Bolivian soil near the Argentine border and that local Bolivian authorities had requested aid from the central government, which, in turn, had sought assistance from the U.S. National Aeronautics and Space Administration to investigate the case.

"A La Paz newspaper said today that there is great interest in learning about the nature of the fallen object, adding that local authorities for security reasons had cordoned off 200 km around the spot where the object fell. The object is said to be a mechanical device with a diameter of almost 4 meters which has already been brought to Tarija. There is interest in determining the accuracy of these reports which have spread quickly throughout the continent, particularly in Bolivia and its neighboring countries. Is it a satellite, a meteorite, or a false alarm?"

151

On May 18, 1978, the U.S. Embassy in La Paz again forwarded a telegram to both NASA and the Secretary of State, Washington, DC. Classified "secret," the telegram disclosed the following:

"Preliminary information provided has been checked with appropriate government agencies. No direct correlation with known space objects that may have re-entered the Earth's atmosphere near May 6 can be made; however, we continue to examine any possibilities. Your attention is invited to State Airgram A-6343, July 26, 1973, which provided background information and guidance for dealing with space objects that have been found. In particular any information pertaining to the pre-impact observations, direction of trajectory, number of objects observed, time of impact and a detailed description including any markings would be useful."

Six days later, a communication was transmitted from the U.S. Defense Attaché Office in La Paz to a variety of U.S. military and government agencies and departments, including NORAD, the Air Force, and the Department of State:

"This office has tried to verify the stories put forth in the local press. The Chief of Staff of the Bolivian Air Force told DATT/AIRA [Defense Attaché/Air Attaché] this date that planes from the BAF have flown over the area where the object was supposed to have landed and in their search they drew a blank. Additionally, DATT/AIRA talked this date with the Commander of the Bolivian Army and informed DATT that the Army's search party directed to go into the area to find the object had found nothing. The Army has concluded that there may or [may] not be an object, but to date nothing has been found."

The CIA's report of May 15, 1978, clearly stated that the object had fallen to earth on Taire Mountain, Bolivia, and had "already been located by authorities." Furthermore, on the following day, the CIA learned that

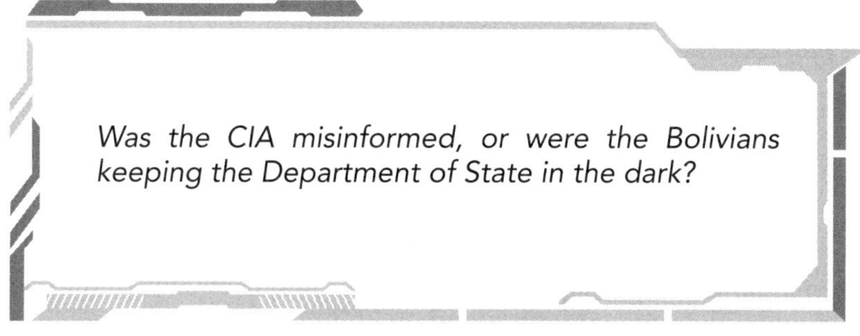
Was the CIA misinformed, or were the Bolivians keeping the Department of State in the dark?

the object had "been brought to Tarija." In contrast, the Bolivian Army and Air Force advised the U.S. Defense Attaché Office that their search for the mysterious object had drawn a blank, and nothing was found. Was the CIA misinformed, or were the Bolivians keeping the Department of State in the dark? Regardless of the answers to those questions, the story does not end there.

Tantalizing information suggests that the object was indeed recovered by U.S. authorities (or at least with the assistance of U.S. authorities) and that both the CIA and NASA played key roles in the case. In June 1979, researcher Leonard Stringfield was contacted by an Argentinean investigator, Nicholas Ojeda, who had interesting data to impart concerning the crash. According to Ojeda: "There is a report of a group of investigators who vanished mysteriously in the area. I really think something big happened in Salta. NASA investigated, but there was no news of it. I have to tell you that in La Paz, Bolivia, a huge Hercules C-130 carried 'something' from the area where the UFO crashed."

In addition, Stringfield's research led to a disclosure that a CIA source known to the researcher Bob Barry confirmed that the C-130 flight took place and that he was aboard the aircraft. "No comment" was the reply that Barry received when the issue of the aircraft's cargo was raised. The available evidence still did not answer the crucial question: What was it that crashed at Bermejo, Bolivia, in May 1978? The possibility that the object was a meteorite almost certainly can be discounted, primarily because of the description given in the Department of State's telegram of May 15: "The object was . . . egg-shaped, metal and about four meters in diameter."

What of the possibility that the object was a man-made satellite? That theory was thrown into doubt by the Department of State's "secret" telegram of 18 May, which was shared with NASA and clearly stated: "No direct correlation with known space objects that may have entered the Earth's atmosphere near May 6 can be made."

In this light, the CIA statement that "people in San Luis and Mendoza provinces reported seeing a flying saucer squadron flying in formation" and the revelations of the Department of State that "the general region has had more than its share of reports of UFOs in the past week" must be considered significant.

Certainly, the most significant development in this case occurred in 2015, when data compiled by the late UFO researcher Bob Pratt surfaced. Pratt was an acclaimed UFO researcher who died in November 2005 and gave me permission to use his material when I was digging into the NASA–UFO controversy. The well-known Brazilian UFO investigator A. J. Gevaerd wrote on November 23, 2005:

"This is a very sad note. Bob Pratt, a great UFO researcher and a good friend, died last Saturday, November 19th. His death is a great loss for ufology worldwide. Last Friday, Bob had a heart attack and by Saturday, he was gone. There will be a memorial service on Friday afternoon at his town, Lake Worth, Florida. Bob was a distinguished American writer, UFO researcher and journalist who was co-author of *Night Siege: The Hudson Valley UFO Sightings*, with J. Allen Hynek, and the author of *UFO Danger in Brazil*, translated and published here as *Perigo Alienígena no Brasil*. It is very probable that no other foreign UFO researcher had more knowledge about the Brazilian ufology as Bob Pratt. He had been to Brazil 13 times and was the kind of field investigator that really went to where the facts were to be researched. He helped a lot the Brazilian UFO researchers to best evaluate the dramatic incidents of alien attacks in the Northeast of the country and in the Amazon. Bob's interest in UFO phenomena began when working for the *National Enquirer*, and was sent to many countries to investigate UFO sightings and ET contacts, such as Argentina, Bolivia, Canada, Chile, Japan, Mexico, Peru and Puerto Rico. Since 1975, he interviewed over two-thousand witnesses, many of them in Brazil. Bob wrote articles for several UFO and non-UFO magazines. He was, for several years, the editor of the renowned *MUFON UFO Journal*."

Of the many people with whom Pratt spoke while on site in Bolivia shortly after the event occurred, one was a physicist, a Dr. Orlando René Bravo. He told Pratt: "People saw different things as the object passed over. In Rosillas, people saw a silver-colored tube or cylinder with a black head in front and flames at the back, a cylinder that appeared to be about four meters long. A teacher in Rosillas told me she saw a fireball pass in the sky and disappear, leaving a trail of smoke behind it, and about five minutes later she heard an explosion. All the teachers and children saw something fall. For twenty to twenty-five miles around, people heard the explosion. I went to all these places, and people said they felt the ground tremble when they heard the explosion."

Dr. Bravo had much more to say, too, including the fact that some witnesses saw two objects: "I don't know what the first object was, but I'm sure the others were missiles. Two geologists from GEOBOL (the Bolivian geological agency) and their guide were in Yerba Buena at the bottom of a ravine, and they thought this object was going to crash on the far side of the hill.

"I walked from La Mamora, about thirty-eight hundred feet high, to Rio Condada, to Puesta de la Laguna, Estancia Jalanoquero and Yerba Buena, which is more than ten thousand feet high. I also walked to the towns of San Luis, Tolomosita, Tolomosa and Pampa Redonda. I interviewed more than fifty people, taking directions with a compass all the time.

"The second investigation was between May 16 and May 21, with Mr. De La Torre. From Cañas up to Mecoya, we interviewed more than thirty people. From Mecoya, we explored up to Cerro Salle, all the way up to border marker number four, at an elevation of nearly twelve thousand feet. I carried a compass, an altimeter and a radiation detector.

"Most of the people in the Mecoya area said the object went to Cerro Bravo in [Colombia]. A sheep-herder said the object exploded in the air near Mecoya and changed direction from the southwest to a more southerly direction.

Cerro Bravo is a stratovolcano in Colombia.

"Apparently, the object crashed into a buttress of Cerro Bravo at about ten thousand feet in height. There, a rockslide can be seen superimposed on the top of an older, natural slide. The difference is clear and can be noticed by the different coloration of the rock. In summing up, we have a complicated problem. One large object came from Sucre to Tarija and changed course, and other objects more or less at the same time came from Emborozu, Palca and Zaire (all southeast of Cerro Bravo).

"The first object's form can't be determined but everybody said it was more or less long, but the others were long and thin like a pencil with a pointed nose and spitting fire from the back. They are maneuverable. They can change directions and they can rise. That's the truth. One of the objects—I'm not sure which—crashed into the mountain and produced an explosion that was heard in La Mamora, Padcaya, Cañas, Camacho and up to Oran in Argentina."

A woman named Guillermina de Antelo, who was in Tarija on the day in question, told Pratt a notable story, too: "I saw a round object like a disc with lights coming out of the back, like fireworks. They were very bright colors, plenty of colors, and most of them were startling pink and yellow. It shocked me. They weren't flames. They were more like beams of light. The object itself was in front and the lights streamed out behind. The object was about the size of my hand in the sky and was round like a disc.

"At first, it seemed slow but then it was very fast and I thought it was going to crash. It was very beautiful. It was like a record or plate from the bottom or edge. I saw it as completely round and I think that when other people say they thought it was long, it was because of the rays of light coming back. It was trailing the rays of light behind, maybe three times as long as the disc itself. The rays were sort of coming back to a point, giving it a sort of fish shape."

Bob Pratt's assessment: "Most likely, something did crash into Cerro Bravo, but whether it was a meteor, a missile or a malfunctioning flying saucer, we may never know. People who saw something flying through the sky gave so many differing descriptions that it could have been any of the three. However, the flight characteristics and the different directions the object took seem to rule out a meteor or any missile, except perhaps for a Cruise-type missile, and that is hardly likely. There is no evidence it was a flying saucer, either, although the descriptions of some witnesses—particularly in the Tarija, Padcaya, Rosillas, Cañas and Mecoya areas—certainly indicate it was an object much like those seen elsewhere in the world."

To this day, the true nature of the events on Taire Mountain, Bolivia, in May 1978 remain shrouded in mystery. As does NASA's secret agenda to uncover the full facts surrounding the UFO phenomenon.

Now let's take a look at the rest of Salter's work as it relates to this issue of using robotic, remote-controlled alien spacecraft designed to destroy our whole civilization.

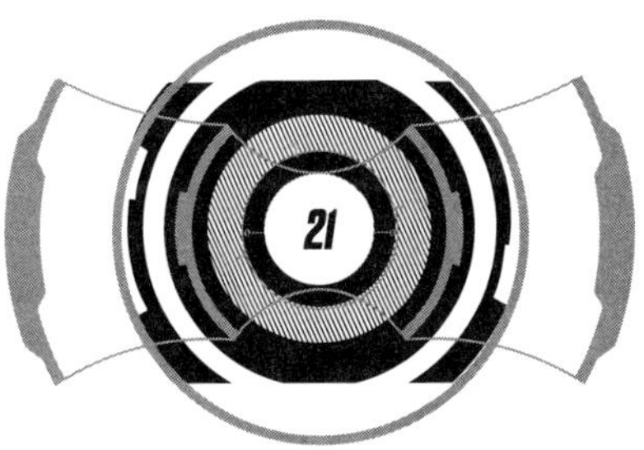

21

From Robots to Viruses

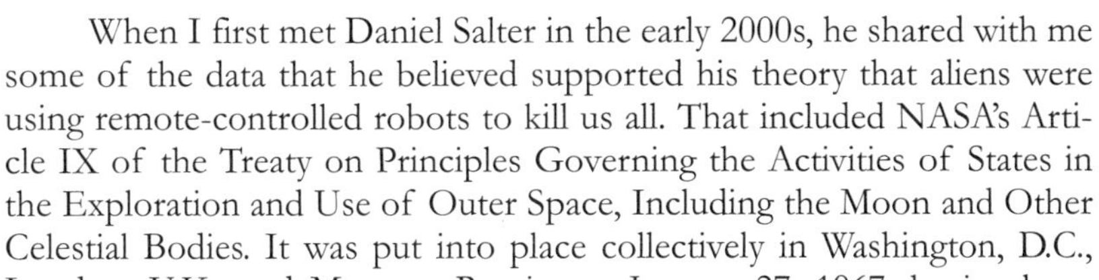

When I first met Daniel Salter in the early 2000s, he shared with me some of the data that he believed supported his theory that aliens were using remote-controlled robots to kill us all. That included NASA's Article IX of the Treaty on Principles Governing the Activities of States in the Exploration and Use of Outer Space, Including the Moon and Other Celestial Bodies. It was put into place collectively in Washington, D.C., London, U.K., and Moscow, Russia, on January 27, 1967, having been created in October of that year. It reads as follows:

"In the exploration and use of outer space, including the Moon and other celestial bodies, States Parties to the Treaty shall be guided by the principle of co-operation and mutual assistance and shall conduct all their activities in outer space, including the Moon and other celestial bodies, with due regard to the corresponding interests of all other States Parties to the Treaty. States Parties to the Treaty shall pursue studies of outer space, including the Moon and other celestial bodies, and conduct exploration of them so as to avoid their harmful contamination and also adverse changes in the environment of the Earth resulting from the introduction of extraterrestrial matter and, where necessary, shall adopt appropriate measures for this purpose.

"If a State Party to the Treaty has reason to believe that an activity or experiment planned by it or its nationals in outer space, including the Moon and other celestial bodies, would cause potentially harmful interference with activities of other States Parties in the peaceful exploration and use of outer space, including the Moon and other celestial bodies, it shall undertake appropriate international consultations before proceeding with any such activity or experiment. A State Party to the Treaty which has reason to believe that an activity or experiment planned by another State Party in outer space, including the Moon and other celestial bodies, would cause potentially harmful interference with activities in the peaceful exploration and use of outer space, including the Moon and other celestial bodies, may request consultation concerning the activity or experiment."

Salter stated that the biggest issue of all this related to what NASA termed: "adverse changes in the environment of the Earth resulting from the introduction of extraterrestrial matter." Right up to his final days, Salter stood by those concerns.

In an article titled "Alien Infection," writer Leslie Mullen stated: "When diseases like SARS, Mad Cow Disease and Monkey-pox cross the species barrier and infect humans, they dominate news headlines. Just imagine, then, the reaction if potentially infectious pathogens were found in rock samples from Mars. As we look toward exploring other worlds, and perhaps even bringing samples back to Earth for testing, astrobiologists have to wonder: could alien pathogens cross the 'planet' barrier and wreak havoc on our world? Even though there is no proof of bacterial or viral pathogens anywhere except Earth, there is already a worried advocacy group called the International Committee Against Martian Sample Return, and science fiction novels like *The Andromeda Strain* depict nightmare alien infection scenarios."

Before his death in 2007, Daniel Salter was a retired agent from the federal government's Interplanetary Phenomenon Scientific and Technical Unit. A retired Air Force pilot, he also worked on Project Blue Book and was an expert on radar.

Another article that Salter shared with me was "Sample Return Missions Scare Some Researchers," an eye-catching story that appeared on Space.com on April 9, 2000. In part, it said: "Researchers, environmentalists and policymakers want NASA to consider carefully its plans to visit and bring back samples from Mars, Europa, and other solar system bodies." Norine Noonan, the director of the Environmental Protection

Agency research arm, came straight to the crux of the worry: "This is a very serious issue, and there are clearly many concerns."

The Space.com article continued: "NASA's longstanding policy is to both protect the Earth and preserve planetary conditions for future biological and organic constituent exploration…. In recent years, NASA has commissioned the National Research Council to conduct several studies on the potential for contaminating other worlds. Now that the space agency also is considering a variety of sample return missions, researchers are becoming increasingly aware that they must put clear standards in place to protect against the highly unlikely but possible introduction—and escape—of extraterrestrial life to Earth. While minimal, the risk level 'is not zero,' warns one NRC study."

Salter had more to provide: a report from writer Robert Roy Britt dated November 27, 2000, stating: "A group of scientists says it has collected an alien bacterium 10 miles above Earth, plus signatures of other extraterrestrial microbes even higher in the atmosphere…. The bacterium was collected at that altitude by a balloon operated by the Indian Space Research Organization. Chandra Wickramasinghe, who leads a study into the results, called the microbe a previously unknown strain of bacteria and said it likely came from a comet. Wickramasinghe and a colleague, Fred Hoyle, say the findings support an idea they pioneered, called panspermia, which holds that the seeds of life are everywhere in space and are the source for life on Earth."

National Geographic News got involved in 2003: "In a letter to the British medical journal *The Lancet*, Chandra Wickramasinghe, from Cardiff University in Wales, and other scientists, propose that SARS may have originated in outer space then fallen down to Earth and landed in China, where the outbreak began. It sounds like a headline from a supermarket tabloid, but the idea may not be as outlandish as it first appears. One hundred tons (90 metric tons) of space debris fall on Earth every day; some scientists believe as much as one ton

Sri Lankan by birth, Chandra Wickramasinghe is a British astronomer and astrobiologist who collaborated for decades with the late astronomer Sir Fred Hoyle.

(0.9 metric ton) of bacteria from space is part of that daily deposit. Particles carrying the SARS virus could have come from a comet, the researchers say, and released into the debris trail of the comet's tail. The Earth's passage through the stream would have led to the entry of the culprit particles. 'We're not saying this is definitely what happened,' said Wickramasinghe, who is also the director of the Cardiff Center for Astrobiology, a research effort that seeks evidence of extraterrestrial life. 'But the theory should not be ruled out.'"

Dr. Robert Wood, who has deeply studied the issue of dangerous, alien viruses, states: "One of those that recognized the potential threat posed by lethal viruses of exotic origins [was] Joshua Lederberg, who was born in New Jersey in 1925, and who obtained his B.A. with honors in Zoology at Columbia. In 1947, Lederberg was appointed Assistant Professor of Genetics at the University of Wisconsin, where he was promoted to Associate Professor in 1950, and subsequently to Professor in 1954. Stanford University Medical School entrusted to him the organization of its Department of Genetics and appointed him Professor and Executive Head in 1959. Lederberg's lifelong research, for which he received the Nobel Prize in 1958 at the age of thirty-three, [was] in genetic structure and function in micro-organisms, and he [was] actively involved in artificial intelligence research, in computer science, and in NASA's experimental programs seeking life on Mars." Lederberg died in 2008.

It's worth noting that Salter was a major supporter of Dr. Wood's work in the field of alien viruses. With that said, now take a look at a 1960s article penned by Lederberg in 1960 with the title "Exobiology: Approaches to Life beyond the Earth." It states in part: "The introduction of microbial life to a previously barren planet, or to one occupied by a less well-adapted form of life, could result in the explosive growth of the implant. With a generation time of 30 minutes and easy dissemination by winds and currents, common bacteria could occupy a nutrient medium the size of the earth in a few days or weeks."

A winner of the 1958 Nobel Prize in Physiology or Medicine, Joshua Lederberg was a molecular biologist with an interest in artificial intelligence.

Lederberg did not end there: "The most dramatic hazard would be the introduction of a new disease, imperiling human health. What we know of the biology of infection makes this an extremely unlikely possibility.... However, a converse argument can also be made, that we have evolved our specific defenses against terrestrial bacteria and that we might be less capable of coping with organisms that lack the proteins and carbohydrates by which they could be recognized as foreign.... At present the prospects for treating a returning vehicle to neutralize any possible hazard are at best marginal by comparison with the immensity of the risk."

Lederberg followed up with yet another article, "A Treaty Proposal on Germ Warfare," in which he states: "The United States has vehemently denied the military use of any biological weapons or of any lethal chemical weapons," adding that research on these weapons nonetheless continued through and after World War II. Lederberg explains, "The Army has a well-known research facility at Fort Detrick, Md., and a testing station at Dugway, Utah.... The large scale deployment of infectious agents is a potential threat against the whole species: mutant forms of viruses could well develop that would spread over the earth's population for a new Black Death.... The future of the species is very much bound up with the control of these weapons."

"Contamination of Mars," a paper from 1967, expands further: "One serious contingency for release of contained micro-organisms is a crash-landing, and, particularly, a spacecraft impacting Mars with a velocity about 6 km/sec will be totally pulverized. But experience with missile impact indicates that even at impact velocities of 0.6 km/sec or less, a significant fraction of the missile's mass is not in the impact crater and is unrecoverable. The shells and grenades used in bacteriological warfare indicate that contained microorganisms will survive such impacts.... In the case of a hard landing, and particularly, of a hypervelocity impact of a spacecraft on Mars, spacecraft fragments will be distributed over a wide area. In a period of minutes—less than the time for many unshielded terrestrial organisms to accumulate an ultraviolet mean lethal dose on Mars—fragments travelling in parabolic trajectories at 6 km/sec will cover a lateral distance ~1000 km."

Then, in September 1967, the *Washington Post* gave Lederberg significant page-space. On the 7th of that month the newspaper ran Lederberg's "Mankind Had a Near Miss from a Mystery Pandemic." His words offered significant food for thought: "In the aftermath of the six-day war in the Middle East last summer, direct air transport from Uganda to Germany and Yugoslavia was disrupted. Shipments of 'green monkeys,' for use in preparing vaccines, were diverted to London airport before transhipment. In the process, a group of at least eight monkeys acquired a disease heretofore unknown to medical science. The disease remains unnamed but might be called Marburg-virus, for it infected at least thirty-two people and killed five of them in Marburg, Germany, and infected two in Frankfurt."

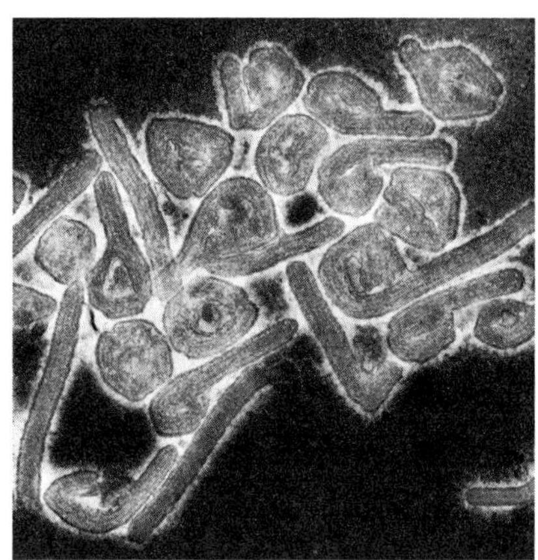

The extremely dangerous Marburg-virus causes hemorrhagic fever in humans. No one knows what the origins of this virus are.

Lederberg went on: "The origin of Marburg-virus is unknown. The threat of a major virus epidemic—a global pandemic—hangs over the head of the species at any time. We were lucky on this occasion, but it was a near miss. It could easily have established a very large focus of infection in countries like India or China or South Vietnam, and in our present knowledge of virology we would have been ill-equipped to stop it from dominating the earth, with a half-billion casualties."

All of this inspired Salter's research and his belief that robotics and viruses might one day turn out to be our downfall. Salter thought that such a thing could happen, too. There was another aspect to all of this, also. Salter was sure that those so-called robotic spheres and tripod devices played a significant role in the matter of so-called cattle mutilations. It's now that the story reaches its pinnacle and becomes even more controversial and dark. For decades, ufologists and conspiracy theorists have pondered on what the purposes of the mutilations are. In light of what we know of Dan Salter's research, it fits in seamlessly with the mutilation phenomenon. Salter was sure that robot drone-like craft were injecting cows—across the United States—with potentially dangerous viruses. On top of that, it's a fact that at many mutilation sites, tripod markings are found—just like the tripod marks that played a role in Salter's research into those mysterious robotic spheres.

The next chapter is a study of the cattle mutilation enigma. Read it carefully, and you'll see how it and Salter's concern blend together.

Trying to Wipe Out the Human Race

Since at least 1967, reports have surfaced throughout the United States of animals, chiefly cattle, slaughtered in bizarre fashion. Organs are taken, and a significant amount of blood is found to be missing. In some cases, the limbs of the cattle are broken, suggesting they have been dropped to the ground from a significant height. Evidence of extreme heat used to slice into the skin of the animals has been found at mutilation sites. Eyes are removed, tongues are sliced off, and typically the sexual organs are gone.

While the answers to the puzzle remain frustratingly outside of the public arena, theories abound. They include extraterrestrials engaged in nightmarish experimentation of the genetic kind; military programs involving the testing of new bio-warfare weapons; occult-based groups that sacrifice the cattle in ritualistic fashion; and government agencies secretly monitoring the food chain, fearful that something worse than "mad cow disease" may have infected the U.S. cattle herd and possibly, as a result, the human population.

Daniel Salter had a deep interest in the cattle mutilation issue, particularly from the 1970s onward, and felt it was tied directly to the matter of aliens trying to eliminate humans by using robotic spheres to dispense alien viruses. He made connections between the "tripod legs" that were seen on many of the robotic spheres mentioned in earlier chapters and identical tripod marks found at the sites of cattle mutilations. Salter was

Tripod marks near where mutilated cattle were found might indicate the presence of extraterrestrials, according to Daniel Salter, who believed aliens were trying to infect livestock with viruses.

sure that the aliens were trying to infect the U.S. cattle herd with diseases that would then infect the human population, too.

Read on and you'll see just how plausible all of Salter's concerns about robotics, viruses, and cattle mutilations are—and how they come together.

From January to March 1973, the state of Iowa was hit hard by cattle mutilations. Not only that, many of the ranchers who lost animals reported seeing strange lights and black-colored helicopters in the direct vicinities of the attacks. That the FBI took keen notice of all this is demonstrated by the fact that, as the Freedom of Information Act has shown, it collected and filed numerous media reports on the cattle mutilations in Iowa. The next piece of data dates from early September 1974. That's when the FBI's director, Clarence M. Kelley, was contacted by Senator Carl T. Curtis, who wished to inform the bureau of a wave of baffling attacks on livestock in Nebraska, the state Curtis represented.

At the time, the FBI declined to get involved, as Director Kelley informed the senator: "It appears that no Federal Law within the investigative jurisdiction of the FBI has been violated, inasmuch as there is no indication of interstate transportation of the maimed animals."

One year later, in August 1975, Senator Floyd K. Haskell of Colorado made his voice known to the FBI on the growing cattle mutilation controversy, stating: "For several months my office has been receiving reports of cattle mutilations throughout Colorado and other western states. At least 130 cases in Colorado alone have been reported to local officials and the Colorado Bureau of Investigation (CBI); the CBI has verified that the incidents have occurred for the last two years in nine states. The ranchers and rural residents of Colorado are concerned and frightened by these incidents. The bizarre mutilations are frightening in themselves: in virtually all the cases, the left ear, rectum and sex organ of each animal has been cut away and the blood drained from the carcass, but with no traces of blood left on the ground and no footprints."

The senator had much more to say, too:

"In Colorado's Morgan County area there has [sic] also been reports that a helicopter was used by those who mutilated the carcasses of the cattle, and several persons have reported being chased by a similar helicopter. Because I am gravely concerned by this situation, I am asking that the Federal Bureau of Investigation enter the case.

Senator Carl Curtis of Nebraska asked the FBI to investigate the mutilations of livestock in his native Nebraska.

"Although the CBI has been investigating the incidents, and local officials also have been involved, the lack of a central unified direction has frustrated the investigation. It seems to have progressed little, except for the recognition at long last that the incidents must be taken seriously. Now it appears that ranchers are arming themselves to protect their livestock, as well as their families and themselves, because they are frustrated by the unsuccessful investigation. Clearly something must be done before someone gets hurt."

Again, the FBI—rather suspiciously, in the opinion of some ranchers and media people—declined to get involved in the investigation of the phenomenon. It was a stance the FBI rigidly stuck to (despite collecting numerous, nationwide newspaper and magazine articles on the subject) until 1978. That was when the FBI learned of an astonishing number of horse and cattle mutilations in Rio Arriba County, New Mexico—mutilations that actually dated back to 1976. They had all been scrupulously investigated and documented by Police Officer Gabe Valdez of Espanola.

It was when the FBI was contacted by New Mexico senator Harrison Schmitt (also the twelfth person to set foot on the moon—in December 1972), who implored the FBI to get involved, that action was finally taken. In March 1979, Assistant Attorney General Philip Heymann prepared a summary on the New Mexico cases for the FBI and, for good measure, photocopied all of Officer Valdez's files for the bureau's director. Things were about to be taken to a new level.

As Valdez's voluminous records showed, from the summer of 1975 to the early fall of 1978, no fewer than 28 cattle mutilation incidents occurred in Rio Arriba County. One of the most bizarre events occurred in June 1976, as Valdez's files demonstrate:

"Investigations around the area revealed that a suspected aircraft of some type had landed twice, leaving three pod marks positioned in a triangular shape. The diameter of each pod was 14 inches. Emanating from the two landings were smaller triangular shaped tripods 28 inches and 4 inches in diameter. Investigation at the scene showed that these small tripods had followed the cow for approximately 600 feet. Tracks of the cow showed where she had struggled and fallen. The small tripod tracks were all around the cow. Other evidence showed that grass around the tripods, as they followed the cow, had been scorched. Also, a yellow oily substance was located in two places under the small tripods. This substance was submitted to the State Police Lab. The Lab was unable to detect the content of the substance.

"A sample of the substance was submitted to a private lab, and they were unable to analyze the substance due to the fact that it disappeared or disintegrated. Skin samples were analyzed by the State Police Lab and the Medical Examiner's Office. It was reported that the skin had been cut with a sharp instrument."

Seventy-two hours later, Valdez liaised with Dr. Howard Burgess of the New Mexico–based Sandia Laboratories, with a view to having the area checked for radiation. It was a wise move. The radiation level was double that which would normally be expected. Valdez's conclusions on this issue: "It is the opinion of this writer that radiation findings are deliberately being left at the scene to confuse investigators."

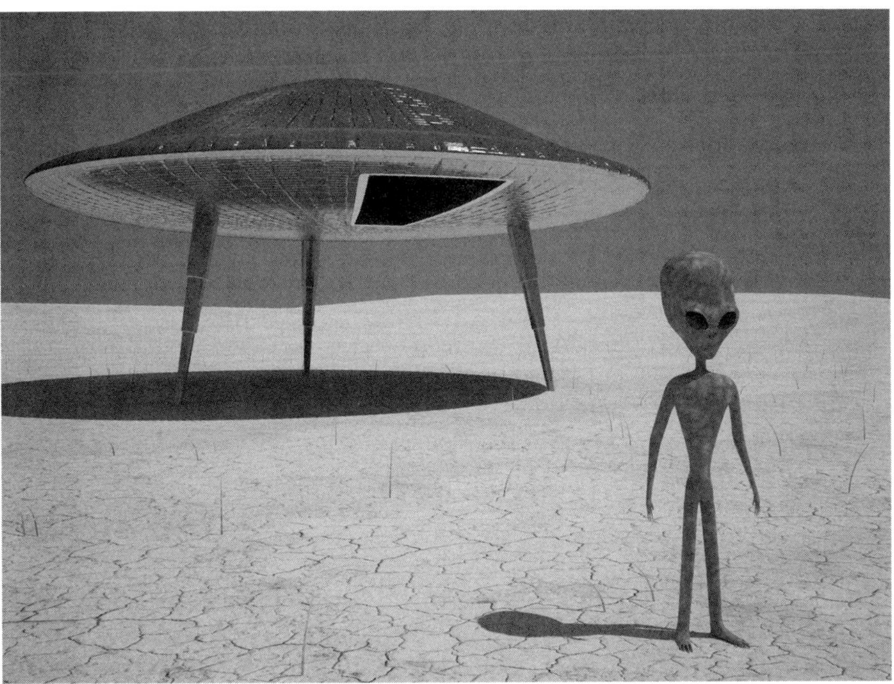

Tripod marks near cattle mutilations seem to be evidence that some type of craft landed near the incidents, possibly something of an extraterrestrial nature.

The case was not over, however. Whatever, or whoever, was responsible for the mutilation made a return visit. Once again, we need to take a look at the official files on the affair. In Valdez's official words: "There was also evidence that the tripod marks had returned and removed the left ear. Tripod marks were found over Mr. Gomez's tire tracks of his original visit. The left ear was intact when Mr. Gomez first found the cow. The cow had a 3-month-old calf which has not been located since the incident. This appears strange since a small calf normally stays around the mother even though the cow is dead."

On the matter of whether the mutilations were the work of cults or natural predators, Valdez said: "Both have been ruled out due to expertise and preciseness and the cost involved to conduct such a sophisticated and secretive operation. It should also be noted that during the spring of 1974 when a tremendous amount of cattle were lost due to heavy snowfalls, the carcasses had been eaten by predators. These carcasses did not resemble the carcasses of the mutilated cows. Investigation has narrowed down to these theories which involve (1) Experimental use of Vitamin B12 and (2) The testing of the lymph node system. During this investigation an intensive study has been made of (3) What is involved in germ warfare testing, and the possible correlation of these 3 factors (germ warfare testing, use of Vitamin B12, testing of the lymph node system)."

A further, very strange, report can be found in Valdez's files, from 1978: "This four-year-old cross Hereford and Black Angus native cow was found lying on left side with rectum, sex organs, tongue, and ears removed. Pinkish blood from [illegible] was visible, and after two days the blood still had not coagulated. Left front and left rear leg were pulled out of their sockets apparently from the weight of the cow which indicates that it was lifted and dropped back to the ground. The ground around and under the cow was soft and showed indentations where the cow had been dropped. 600 yards away from the cow were the 4-inch circular indentations similar to the ones found at the Manuel Gomez ranch on 4-24-78.

"This cow had been dead approximately [illegible] hours and was too decomposed to extract samples. This is the first in a series of mutilations in which the cows' legs are broken. Previously the animals had been lifted from the brisket with a strap. These mutilated animals all dehydrate rapidly (in one or two days)."

As the summer of 1978 progressed, so did the number of reports where elevated radiation readings were found, as Valdez noted in his records: "It is believed that this type of radiation is not harmful to humans, although approximately 7 people who visited the mutilation site complained of nausea and headaches. However, this writer has had no such symptoms after checking approximately 11 mutilations in the past 4 months. Identical mutilations have been taking place all over the Southwest. It is strange that no eyewitnesses have come forward or that no accidents [have] occurred. One has to admit that whoever is responsible for the mutilations is very well organized with boundless financing and secrecy. Writer is presently getting equipment through the efforts of Mr. Howard Burgess, Albuquerque, N.M. to detect substances on the cattle which might mark them and be picked up by infra-red rays but not visible to the naked eye."

A lengthy document, prepared by Forrest S. Putman, the FBI's Special-Agent-in-Charge at Albuquerque,

Elevated radiation readings are commonly detected in areas where cattle have been found mutilated.

New Mexico, was soon thereafter sent to the FBI's headquarters in Washington, D.C. It read:

"Information furnished to this office by Officer Valdez indicates that the animals are being shot with some type of paralyzing drug and the blood is being drawn from the animal after an injection of an anti-coagulant. It appears that in some instances the cattle's legs have been broken and helicopters without any identifying numbers have reportedly been seen in the vicinity of these mutilations.

"Officer Valdez theorizes that clamps are being placed on the cow's legs and they are being lifted by helicopter to some remote area where the mutilations are taking place and then the animal is returned to its original pasture. The mutilations primarily consist of removal of the tongue, the lymph gland, lower lip and the sexual organs of the animal.

"Much mystery has surrounded these mutilations, but according to witnesses they give the appearance of being very professionally done with a surgical instrument, and according to Valdez, as the years progress, each surgical procedure appears to be more professional. Officer Valdez has advised that in no instance, to his knowledge, are these carcasses ever attacked by predator or scavenger animals, although there are tracks which would indicate that coyotes have been circling the carcass from a distance. Special Agent Putman then informed the Director of the outcome of Valdez's run-ins with officials.

"He also advised that he has requested Los Alamos Scientific Laboratory to conduct investigation for him but until just recently has always been advised that the mutilations were done by predatory animals. Officer Valdez stated that just recently he has been told by two assistants at Los Alamos Scientific Laboratory that they were able to determine the type of tranquilizer and blood anti-coagulant that have been utilized."

Putnam then demonstrated to headquarters the astonishing scale of the mutilation puzzle: "Officer Valdez stated that Colorado probably has the most mutilations occurring within their State and that over the past four years approximately 30 have occurred in New Mexico. He stated that of these 30, 15 have occurred on Indian Reservations but he did know that many mutilations have gone unreported which have occurred on the Indian reservations because the Indians, particularly in the Pueblos, are extremely superstitious and will not even allow officers in to investigate in some instances. Officer Valdez stated since the outset of these mutilations there have been an estimated 8,000 animals mutilated which would placc the loss at approximately $1,000,000."

Putman additionally advised the director: "It is obvious if mutilations are to be solved there is a need for a coordinated effort so that all material available can be gathered and analyzed and further efforts synchronized. Whether the FBI should assume this role is a matter to be decided. If we are merely to investigate and direct our efforts toward the 15 mutilated cattle on the Indian reservation, we, I believe, will be in the same position as the other law enforcement agencies at this time and would be seeking to achieve an almost impossible task.

"It is my belief that if we are to participate in any manner that we should do so fully, although this office and the USA's office are at a loss to determine what statute our investigative jurisdiction would be in this matter. If we are to act solely as a coordinator or in any other official capacity the sooner we can place this information in the computer bank, the better off we would be and in this regard it would be my recommendation that an expert in the computer field at the Bureau travel to Albuquerque in the very near future so that we can determine what type of information will be needed so that when the invitation for the April conference is submitted from Senator Schmitt's Office that the surrounding States will be aware of the information that is needed to place in the computer.

Senator Harrison Schmitt of New Mexico, a former NASA astronaut, was one of several state leaders concerned about the continued trend of livestock mutilations.

"It should be noted that Senator Schmitt's Office is coordinating the April conference and will submit the appropriate invitations and with the cooperation of the USA, Mr. Thompson will chair this conference. The FBI will act only as a participant."

Putnam went on to describe the theories that had been advanced to try and explain the phenomenon: "Since this has not been investigated by the FBI in any manner we have no theories whatsoever as to why or what is responsible for these cattle mutilations. Officer Gabe Valdez is very adamant in his opinion that these mutilations are the work of the U.S.

government and that it is some clandestine operation either by the CIA or the Department of Energy and in all probability is connected with some type of research into biological warfare. His main reason for these beliefs is that he feels that he was given the 'run around' by Los Alamos Scientific Laboratory and they are attempting to cover up this situation. There are also theories that these are cults (religious) of some type of Indian rituals resulting in these mutilations and the wildest theory advanced is that they have some connection with unidentified flying objects."

In the closing section of his report, Putman said: "If we are to assume an investigative posture into this area, the matter of manpower, of course, becomes a consideration and I am unable to determine at this time the amount of manpower that would be needed to give this our full attention so that a rapid conclusion could be reached. The Bureau is requested to furnish its comments and guidance on this whole situation including, if desired, the Legal Counsel's assessment of jurisdictional question. An early response would be needed, however, so that we might properly, if requested to do so, obtain the data bank information. If it appears that we are going to become involved in this matter, it is obvious that there would be a large amount of correspondence necessary and Albuquerque would suggest a code name be established of BOVMUT."

As a result of the growing concern surrounding the cattle mutilations, a conference on the subject was held on April 20, 1979, at the Albuquerque Public Library. There was a heavy concentration of FBI agents at the conference....

As a result of the growing concern surrounding the cattle mutilations, a conference on the subject was held on April 20, 1979, at the Albuquerque Public Library. There was a heavy concentration of FBI agents at the conference, which resulted in the following official document being prepared to summarize the various theories, cases, and ideas advanced at the conference:

"Forrest S. Putman, Special Agent in Charge (SAC), Albuquerque Office of the FBI, explained to the conference that the Justice Department had given the FBI authority to investigate those cattle mutilations which have occurred or might occur on Indian lands. He further explained that the Albuquerque FBI would look at such mutilations in connection with mutilations occurring off Indian lands for the purpose of comparison and control, especially where the same methods of operation are noted. SAC Putman said that in order for this matter to be resolved, the facts surrounding such mutilations should be gathered and computerized.

"District Attorney Eloy Martinez, Santa Fe, New Mexico, told the conference that his judicial district had made application for a $50,000 Law Enforcement Assistance Administration (LEAA) Grant for the purpose of investigating the cattle mutilations. He explained that there is hope that with the funds from this grant, an investigative unit can be established for the sole purpose of resolving the mutilation problem. He said it is his view that such an investigative unit could serve as a headquarters for all law enforcement officials investigating the mutilations and, in particular, would serve as a repository for information developed in order that this information could be coordinated properly. He said such a unit would not only coordinate this information, but also handle submissions to a qualified lab for both evidence and photographs. Mr. Martinez said a hearing will be held on April 24, 1979, for the purpose of determining whether this grant will be approved.

"Gabe Valdez, New Mexico State Police, Dulce, New Mexico, reported he has investigated the death of 90 cattle during the past three years, as well as six horses. Officer Valdez said he is convinced that the mutilations of the animals have not been the work of predators because of the precise manner of the cuts. Officer Valdez said he had investigated mutilations of several animals which had occurred on the ranch of Manuel Gomez of Dulce, New Mexico.

"Manuel Gomez addressed the conference and explained he had lost six animals to unexplained deaths which were found in a mutilated condition within the last two years. Further, Gomez said that he and his family are experiencing fear and mental anguish because of the mutilations.

"David Perkins, Director of the Department of Research at Libre School in Farasita, Colorado, exhibited a map of the United States which contained hundreds of colored pins identifying mutilation sites. He commented that he had been making a systematic collection of data since 1975 and has never met a greater challenge. He said, 'The only thing that makes sense about the mutilations is that they make no sense at all.'

"Tom Adams of Paris, Texas, who has been independently examining mutilations for six years, said his investigation has shown that helicopters are almost always observed in the area of the mutilations. He said that the helicopters do not have identifying markings and they fly at abnormal, unsafe, or illegal altitudes.

"Dr. Peter Van Arsdale, Ph.D., Assistant Professor, Department of Anthropology, University of Denver, suggested that those investigating the cattle mutilations take a systematic approach and look at all types of evidence in discounting any of the theories such as responsibility by extraterrestrial visitors or satanic cults.

Peter Van Arsdale, Ph.D., is a retired professor at the University of Denver with a background in psychology and anthropology. He advised the government about the cattle mutilations' possible causes.

"Richard Sigismund, Social Scientist, Boulder, Colorado, presented an argument which advanced the theory that the cattle mutilations are possibly related to activity of UFOs. Numerous other persons made similar type presentations expounding on their theories regarding the possibility that the mutilations are the responsibility of extraterrestrial visitors, members of Satanic cults, or some unknown government agency.

"Dr. Richard Prine, Forensic Veterinarian, Los Alamos Scientific Laboratory (LASL), Los Alamos, New Mexico, discounted the possibility that the mutilations had been done by anything but predators. He said he had examined six carcasses and in his opinion predators were responsible for the mutilation of all six.

"Dr. Claire Hibbs, a representative of the State Veterinary Diagnostic Laboratory, New Mexico State University, Las Cruces, New Mexico, said he recently came to New Mexico, but that prior to that he examined some mutilation findings in Kansas and Nebraska. Dr. Hibbs said the mutilations fell into three categories: animals killed and mutilated by predators and scavengers, animals mutilated after death by 'sharp instruments' and animals mutilated by pranksters.

"Tommy Blann, Lewisville, Texas, told the conference he has been studying UFO activities for twenty-two years and mutilations for twelve years. He explained that animal mutilations date back to the early 1800's in England and Scotland. He also pointed out that animal mutilations are not confined to cattle, but cited incidents of mutilation of horses, dogs, sheep, and rabbits. He also said the mutilations are not only nationwide, but international in scope.

"Chief Raleigh Tafoya, Jicarilla Apache Tribe, and Walter Dasheno, Governor, Santa Clara Pueblo, each spoke briefly to the conference. Both spoke of the cattle which had been found mutilated on their respective Indian lands. Chief Tafoya said some of his people who have lost livestock have been threatened.

"Carl W. Whiteside, Investigator, Colorado Bureau of Investigation, told the conference that between April and December 1975, his Bureau investigated 203 reports of cattle mutilations."

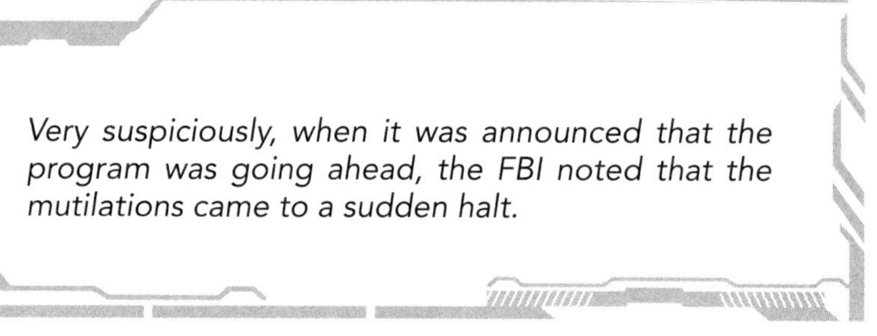
Very suspiciously, when it was announced that the program was going ahead, the FBI noted that the mutilations came to a sudden halt.

One month later, the District Attorney's Office for Santa Fe, New Mexico, secured $50,000 in funding to allow a detailed study of the evidence to commence—specifically in New Mexico. Very suspiciously, when it was announced that the program was going ahead, the FBI noted that the mutilations came to a sudden halt. This gave rise to deep suspicions that the mutilators were all too human and, having heard of the planned investigation, hastily backed away until matters calmed down—which they did, when the number of new reports trailed off to nothing.

With a distinct lack of new data to go on, the ambitious program was left to study a mere handful of cases—all of which it relegated to the work of predators and absolutely nothing else. Hardly surprisingly, this gave rise to the suspicions, among conspiracy theorists, that this was the goal all along: launch an investigation and assert that the mutilations were the work of predators and nothing else, and then close the investigation down. If that was the case, then it worked all too well: the world

of officialdom walked away from the mutilation problem, asserting that it had resolved the entire matter in down-to-earth terms and assuring the public and the ranching community that there was nothing to worry about.

Evidently, however, there *was* something to worry about: no sooner had the project closed down when the mutilations began again. And they continue to this day. Whether the work of the government, the military, satanic cults, or deadly extraterrestrials, all that can be said of the cattle mutilations is that, officially, at least, they are no longer of any interest to the FBI or to any other arm of officialdom.

23

From Robots to Wormholes

On several occasions in the pages of this book I have addressed the matter of robots and aliens—particularly in relation to the story of Philip Corso, biological robots, time travel, and the Roswell UFO event of July 1947. There's one angle that has not been touched upon yet but most assuredly does need addressing. No matter the size and intelligence of robots that travel throughout the universe, they will require a novel way to achieve such a thing. Going directly from point to point simply won't work. So, if biological robots have made it to Earth from another solar system, then they must be using a science that, so far, is way beyond our skills. That science almost certainly revolves around the concept of what are known as wormholes. Let's have a look at what we know about this admittedly weird phenomenon.

Nola Taylor Redd is a writer who has carefully studied the subject of wormholes. As her bio notes: "Nola Taylor Redd is a contributing writer for Space.com. She loves all things space and astronomy-related, and enjoys the opportunity to learn more. She has a Bachelor's degree in English and Astrophysics from Agnes Scott College and served as an intern at *Sky & Telescope* magazine."

Redd says: "The wormhole theory postulates that a theoretical passage through space-time could create shortcuts for long journeys across the universe. Wormholes are predicted by the theory of general relativity. But be wary: wormholes bring with them the dangers of sudden collapse, high radiation and dangerous contact with exotic matter. Wormholes were first theorized in 1916, though that wasn't what they were called at the time. While reviewing another physicist's solution to the equations in Albert Einstein's theory of general relativity, Austrian physicist Ludwig Flamm realized another solution was possible. He described a 'white hole,' a theoretical time reversal of a black hole. Entrances to both black and white holes could be connected by a space-time conduit."

Jillian Scharr is a staff writer for Seattle-based Harebrained Schemes. She, too has, carefully addressed the almost-mind-boggling issues of time-travel and wormholes: "The concept of a time machine typically conjures up images of an implausible plot device used in a few too many science-fiction storylines. But according to Albert Einstein's general theory of relativity, which explains how gravity operates in the universe, real-life time travel isn't just a vague fantasy. Traveling forward in time is an uncontroversial possibility, according to Einstein's theory. In fact, physicists have been able to send tiny particles called muons, which are similar to electrons, forward in time by manipulating the gravity around them. That's not to say the technology for sending humans 100 years into the future will be available anytime soon, though. Time travel to the past, however, is even less understood. Still, astrophysicist Eric W. Davis, of the Earth-Tech International Institute for Advanced Studies at Austin, argues that it's possible. All you need, he says, is a wormhole, which is a theoretical passageway through space-time that is predicted by relativity."

Now let us ponder deeply on the words of NASA media relations specialist Calla Cofield, whose work "has appeared in *APS News*, *Symmetry* magazine, *Scientific American*, *Nature News*, *Physics World*, and others. From 2010 to 2014 she was a producer for the *Physics Central Podcast.* Previously, Calla worked at the American Museum of Natural History in New York City (hands down the best office building ever) and SLAC National Accelerator Laboratory in California." On the matter of wormholes, Cofield says:

Famed mathematician and physicist Albert Einstein theorized that the speed of light was as fast as anyone could travel. Modern-day physicists have been trying to get around that limitation for years.

"In his 1994 publication *Black Holes and Time Warps* . . . [Kip] Thorne proposes a thought experiment: Say he obtains a small wormhole, which connects two points in space as if they were not separated by any distance at all. Thorne takes his wormhole and puts one end in his living room, and the other aboard a spaceship parked in his front yard. Thorne's wife, Carolee, hops aboard the spaceship to prepare for a trip. The two don't have to say goodbye, though, because no matter how far away Coralee travels, they can see each other through the wormhole. They can even hold hands, as if through an open doorway.

"Carolee starts up the spaceship, heads into space and travels for six hours at the speed of light. She then turns around and comes back home traveling at the same speed—a round trip of 12 hours. Thorne watches through the wormhole and sees this trip occur. He sees Coralee return from her trip, land on the front lawn, get out of the spaceship and head into the house. But when Thorne looks out the window in his own world, his front lawn is empty. Coralee has not returned. Because she traveled at the speed of light, time slowed down for her: What was 12 hours for her was 10 years for Thorne back on Earth."

Fascinating? Indeed! We aren't over yet, though.

Bill Andrews of *Discover* magazine most assuredly gives us something to think about: "The first problem for any explorer determined to survey a wormhole is simply finding one. While Einstein's work says they can exist, we don't currently know of any. They may actually be impossible after all, forbidden by some deeper physics that the universe obeys, but we haven't discovered. The second issue is that, despite years of research, scientists still aren't really sure how wormholes would work. Can any technology ever create and manipulate them, or are they simply a part of the universe? Do they stay open forever, or are they only traversable for a limited time? And perhaps most significantly, are they stable enough to allow for human travel? The answer to all of these: We just don't know. But that doesn't mean scientists aren't working on it. Despite the lack of actual wormholes to study, researchers can still model and test Einstein's equations. NASA's conducted legitimate wormhole research for decades, and a team described [in 2020] how wormhole-based travel might be more feasible than previously thought."

Now I'll share with you the words of my colleague and friend at the *Mysterious Universe* website, Paul Seaburn. In 2018, Seaburn elected to immerse himself into this strange and swirling world of wormholes: "In a recent post on his *Forbes* blog, *Starts with a Bang*, theoretical astrophysicist and science writer Ethan Siegel lays out the parts and the plans for traveling backwards in time. Siegel claims this 'time machine' abides by Einstein's general theory of relativity and will not destroy the universe as we know it. Siegel proposes a sort of reverse wormhole. Instead of the conventional 'travel 40 light years out at nearly the speed of light, come back and you've aged 2 years while everyone else is 82 years older,' he proposes a wormhole with one fixed end and one that moves around at nearly the speed of light. The wormhole is created, you wait a year and then enter the end that has been in motion. When you come out at the fixed end, it's 40 years prior. That means if you entered this wormhole today, you could travel back to 1978."

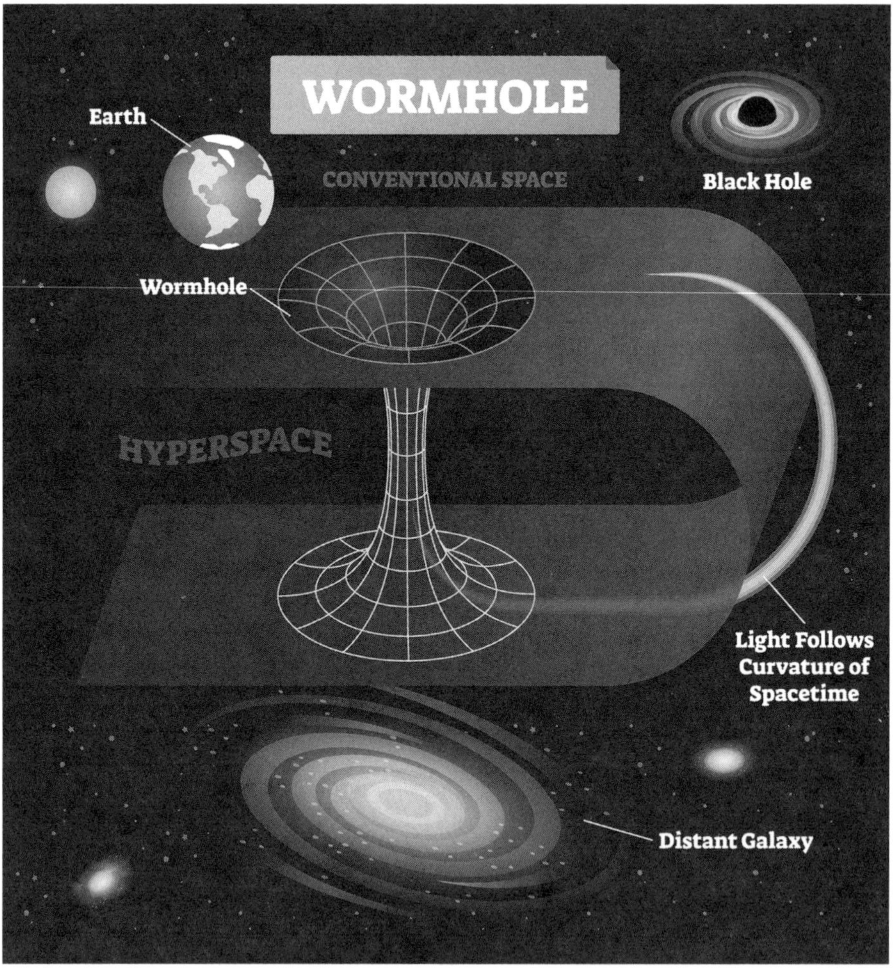

Wormholes (which are still theoretical at this time) are distortions of space-time that form tunnels connecting two distant places in the universe, thus allowing faster-than-light travel.

To conclude this particular aspect of both space travel and time travel, here are important observations from Richard F. Holman, dean at Minerva University and a former professor of physics at Carnegie Mellon University: "Wormholes are solutions to the Einstein field equations for gravity that act as 'tunnels,' connecting points in space-time in such a way that the trip between the points through the wormhole could take much less time than the trip through normal space. The first wormhole-like solutions were found by studying the mathematical solution for black holes. There it was found that the solution lent itself to an extension whose geometric interpretation was that of *two* copies of the black hole geometry connected by a 'throat' (known as an Einstein-Rosen bridge). The throat is a dynamical object attached to the two holes that pinches off extremely quickly into a narrow link between them. Theorists have since found other wormhole solutions; these solutions connect various types of geometry on either mouth of the wormhole. One amazing aspect of wormholes is that because they can behave as 'shortcuts' in space-time, they must allow for backwards time travel! This property goes back to the usual statement that if one could travel faster than light, that would imply that we could communicate with the past."

It's most important to note that all of those who have addressed the issue of wormholes are well-respected figures in their own specific arenas. And, they have brought to the table significant material that suggests wormholes, biological robots, and time travel are a reality—something that is most important when it comes to addressing the data ahead.

Robot Assassins

Over the decades, there have been many high-profile assassinations. Not only that; there is good evidence that at least some of the assassins were subjected to mind control. In other words, those who shot the bullets had little knowledge of or control over their actions. They were, in essence, acting like programmed robots. Thanks to the provisions of the Freedom of Information Act, the CIA has placed in the public domain an early 1950s–era document titled "A Study of Assassination." It's a guide designed to teach CIA agents how to kill by assassination and make it look like something else did it. Surely, it's one of the most unique official CIA documents ever constructed. The unknown agent that worked on the program—whose name is carefully blacked out on the file—says:

"Assassination is a term thought to be derived from 'Hashish,' a drug similar to marijuana, said to have been used by Hasan-i-Sabah to induce motivation in his followers, who were assigned to carry out political and other murders, usually at the cost of their lives. It is here used to describe the planned killing of a person who is not under the legal jurisdiction of the killer, who is not physically in the hands of the killer, who has been selected by a resistance organization for death, and whose death provides positive advantages to that organization.

"Assassination is an extreme measure not normally used in clandestine operations. It should be assumed that it will never be ordered or authorized by any U.S. Headquarters, though the latter may in rare instances agree to its execution by members of an associated foreign service. This reticence is partly due to the necessity for committing communications to paper. No assassination instructions should ever be written or recorded. Consequently, the decision to employ this technique must nearly always be reached in the field, at the area where the act will take place. Decision and instructions should be confined to an absolute minimum of persons. Ideally, only one person will be involved. No report may be made, but usually the act will be properly covered by normal news services, whose output is available to all concerned."

The document continues:

The act of assassinating someone, while not considered morally acceptable, is still used as an extreme measure to protect democracy—or so the logic goes when it comes to the CIA and its "How to Kill" document.

"Murder is not morally justifiable. Self-defense may be argued if the victim has knowledge which may destroy the resistance organization if divulged. Assassination of persons responsible for atrocities or reprisals may be regarded as just punishment. Killing a political leader whose burgeoning career is a clear and present danger to the cause of freedom may be held necessary. But assassination can seldom be employed with a clear conscience. Persons who are morally squeamish should not attempt it.

"The techniques employed will vary according to whether the subject is unaware of his danger, aware but unguarded, or guarded. They will also be affected by whether or not the assassin is to be killed with the subject hereafter, assassinations in which the subject is unaware will be termed 'simple'; those where the subject is aware but unguarded will be termed 'chase'; those where the victim is guarded will be termed 'guarded.' If the assassin is to die with the subject, the act will be called 'lost.'

"If the assassin is to escape, the adjective will be 'safe.' It should be noted that no compromises should exist here. The assassin must not fall alive into enemy hands. A further type [of] division is caused by the need to conceal the fact that the subject was actually the victim of assassination, rather than an accident or natural causes. If such concealment is desirable the operation will be called 'secret'; if concealment is immaterial, the act will be called 'open'; while if the assassination requires publicity to be effective it will be termed 'terroristic.'"

Now we come to the most important—and undeniably relevant—section of the CIA's "How to Kill" document. It highlights the means by which alcohol and medication can be used to make a murder look like an accidental death or a case of planned suicide. The CIA says:

"In all types of assassination except terroristic, drugs can be very effective. If the assassin is trained as a doctor or nurse and the subject is under medical care, this is an easy and rare method. An overdose of morphine administered as a sedative will cause death without disturbance and is difficult to detect. The size of the dose will depend upon whether the subject has been using narcotics regularly. If not, two grains will suffice. If the subject drinks heavily, morphine or a similar narcotic can be injected at the passing out stage, and the cause of death will often be held to be acute alcoholism."

Lee Harvey Oswald, Sirhan Sirhan, and Mark Chapman—to name three infamous assassins—all had "intriguing" backgrounds.

The *Atomic Poet* website notes: "After Mark David Chapman shot and killed John Lennon, he calmly opened up *Catcher in the Rye* and proceeded to read it—before being apprehended. John Hinckley, the man who attempted to kill Ronald Reagan, also was in possession of the book. It is also alleged Lee Harvey Oswald was quite fond of the book, though this is disputed. *Catcher in the Rye* has sold 65 million copies. Of the millions who have enjoyed the book, perhaps three have become well-known assassins. Still, we should ask: is there any merit to the book being an assassination trigger?"

Writer Aidan Doyle says: "There are enough rumors about murders linked to J. D. Salinger's classic that the unwitting assassins in the Mel Gibson film *Conspiracy Theory* are portrayed as being brainwashed with the urge to buy the novel. John Lennon's murderer, Mark David Chapman, was famously obsessed with *The Catcher in the Rye*. Chapman wanted to change his name to Holden Caulfield and once wrote in a copy of the book 'This is my statement,' and signed the protagonist's name. He had a copy of the book in his possession when the police arrested him."

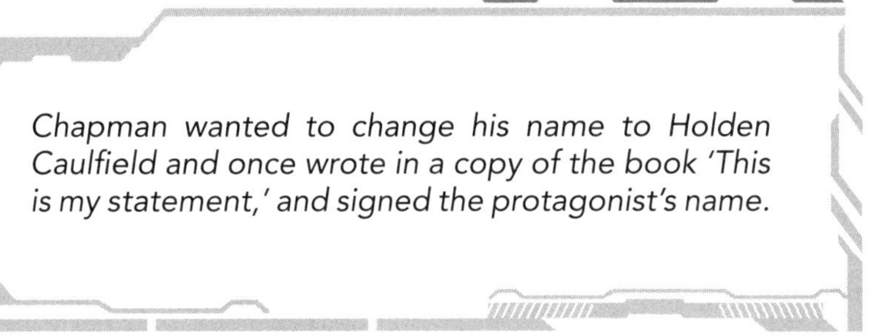

Chapman wanted to change his name to Holden Caulfield and once wrote in a copy of the book 'This is my statement,' and signed the protagonist's name.

But why, exactly, should the book have any bearing whatsoever on Mark Chapman's crazed killing of John Lennon? Conspiracy theorists maintain that trained, mind-controlled assassins—born out of the CIA's controversial MKUltra program—are "switched on" by certain key "trigger words" that appear in the text of *The Catcher in the Rye*. Lawrence Wilson, M.D., notes that a "hypnotist can implant the suggestion that when the phone rings twice, or when the doorbell rings, a post-hypnotic suggestion such as to kill whomever is in the room, even if it is your wife, will go into effect. This is used by some foreign police agencies to train hypnotized assassins." In other words, so the theory goes, Mark Chapman may have been a victim of deep hypnosis on the part of MKUltra operatives, but rather than relying on a phone or doorbell ringing, they used *The Catcher in the Rye*—or segments of the text—as the trigger that turned Chapman into a ruthless killer—one who had no control over his deadly actions on December 8, 1980.

The blog *CIA Killed Lennon* records: "While a teenager in Decatur, Georgia, Chapman did a lot of LSD, then found Jesus, and devoted his life to working with the YMCA, which, according to Philip Agee (*CIA Diary*, 1975), was prime recruiting grounds for CIA stations in Latin America. Chapman's YMCA employment records are missing. In June 1975, Chapman volunteered to work in the YMCA office in Beirut, Lebanon, as the civil war erupted. Returning to the U.S., Chapman was sent to work with newly resettled Vietnamese refugees (and CIA assets) in Fort Chaffee, Arkansas, run by World Vision, an evangelical organization accused of CIA collaboration in Honduras and El Salvador."

Let's not forget, too, that the doorman at the Dakota hotel, Jose Sanjenis Perdomo, had worked with both Cuba's secret police and the CIA. On the matter of Perdomo, *Rumor Mill News* stated in 2004:

"Newly discovered information about doorman Jose Perdomo suggests he may have been John Lennon's true assassin and Mark David Chapman was merely a patsy who confessed to the crime while under the spell of relentless mind control techniques such as hypnosis, drug abuse, shock treatment, sleep deprivation, and so on."

It's now time for us to take a look at the controversial world of mind control in relation to James Earl Ray, who is said to have been the man who shot and killed Martin Luther King Jr. on April 4, 1968. It is a fact that after he escaped from prison in 1967, Ray underwent plastic surgery—reportedly, noted Brad Steiger, on the orders of Raoul. Steiger, who had dug deep into the MLK killing, said: "A link to MK-ULTRA, the CIA's mind-control project, may have occurred when Ray was recuperating from plastic surgery. A Dr. William Joseph Bryan Jr. had programmed individuals when he was with the air force as chief of Medical Survival Training, the air force's covert mind-control section. Bryan, whom some called pompous and arrogant, liked nothing better than to talk about himself and his accomplishments. He was known as an expert on brainwashing, and he served as a consultant on *The Manchurian Candidate*, a motion picture that portrayed a programmed political assassin. In informal discussions, Bryan 'leaked' that he had programmed Sirhan Sirhan and James Earl Ray to commit assassinations and to forget their participation in the act."

Almost two months to the day after Martin Luther King Jr. was shot and killed in Memphis, Tennessee, the life of yet another formidable figure in American history was brought to a violent, bloody, and controversial end. That figure was Robert Francis Kennedy. He was better known as Bobby, as the younger brother of President John F. Kennedy—himself shot and killed under suspicious circumstances in Dallas, Texas, in November 1963—and as the man who served as the attorney general of the United States while JFK was in office. As with the deaths of Martin Luther King Jr. and President Kennedy, the circumstances surrounding the shooting of RFK are steeped in mystery. Similarly, as with the deaths of King and JFK, there is more than one version of events.

The version that is accepted by the U.S. government is that Kennedy was shot and killed in the early hours of June 5, 1968, in Los Angeles, California, by a man named Sirhan Sirhan—and *only* by Sirhan Sirhan. Doubts about Sirhan's guilt surfaced very soon after he was found guilty of killing RFK. While in San Quentin prison in 1969, Sirhan was interviewed by a man named Dr. Eduard Simson-Kallas, an expert in the field of hypnosis, who believed Sirhan had been subjected to some form of subliminal programming of the

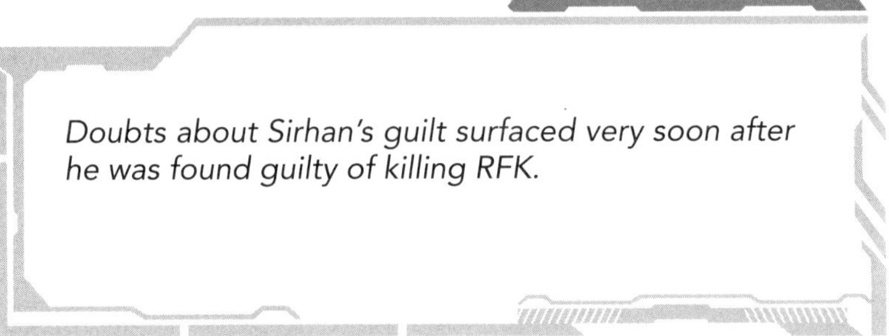
Doubts about Sirhan's guilt surfaced very soon after he was found guilty of killing RFK.

mind-control kind. On top of that, the coroner in the RFK case, Thomas Noguchi, offered his conclusion that the bullet that entered the senator's skull, behind his right ear, had been fired at a distance of barely one inch. That suggests that even if Sirhan was a hypnotically controlled assassin, he must have had an accomplice, since he was most definitely farther away than one inch when he shot Robert Kennedy.

In 1978, 14 years after the Warren Commission laid all the blame for President John Kennedy's assassination firmly on the shoulders of Lee Harvey Oswald, the U.S. House Select Committee on Assassinations came to a different conclusion. The lone gunman, said the committee, was not such a lone gunman after all. President Kennedy's death was the result of nothing less than a full-on conspiracy. The HSCA agreed with the Warren Commission that Kennedy was killed by Oswald and no one else; however, the committee went one step further by concluding that Oswald was not the only gunman prowling around Dallas on that deadly day. Forensic analysis suggested to the HSCA's investigators that *four* shots rang out, not the three that the Warren Commission attributed to Oswald. That is to say, there was another gunman. In the minds of the HSCA's members, this mysterious second character completely missed his target. Nevertheless, a pair of shooters meant a conspiracy was at the heart of the JFK assassination. In other words: take that, Warren Commission.

In October 1959, Lee Harvey Oswald—a self-admitted Marxist—made his way to the Soviet Union. Oswald reached Moscow on October 16 and announced that he wished to remain in Russia. Although the Soviets were initially reluctant to allow Oswald residency, that soon changed. It wasn't long before Oswald had a job and a home. In 1961, he had a wife, Marina. Fatherhood soon followed. Claiming to have become disillusioned with a dull life in the Soviet Union, however, Oswald moved his family to the United States in 1962. Was Oswald recruited by the KGB during his time in Russia? Did his return to the States actually have nothing to do with disillusionment? Had the elite of the Kremlin convinced Oswald to kill Kennedy? One person who has commented on such matters is Ion Mihai Pacepa. In 1978, Pacepa, a general with Romania's Department of

State Security, defected to the United States. One of Pacepa's revelations was that JFK was killed on the orders of Soviet premier Nikita Khrushchev. Still seething from backing down in the Cuban missile crisis of 1962, Khrushchev was determined to exact his revenge. Oswald was chosen to ensure that revenge was achieved. Notably, Pacepa asserted that Khrushchev made a last-minute decision not to go ahead with the plan to kill JFK. Unfortunately, the Russians failed to make timely contact with Oswald and inform him of the change in plans. The countdown to assassination could not be stopped.

What all of these high-profile, history-shattering killings demonstrate is that not a single one of those alleged assassins had what could be called a totally normal background. The issue of mind control and manipulation hung over them at various times. Controlled, brain-washed patsies who had no abilities to stop themselves from killing MLK, RFK, JFK, and John Lennon? That's very much how it looks.

Advanced Research Projects

One of the lesser-known agencies that has long played a significant role in the advancement of computer technology, including artificial intelligence, is the Defense Advanced Research Projects Agency (DARPA). The agency says of its work and history: "For more than fifty years, DARPA has held to a singular and enduring mission: to make pivotal investments in breakthrough technologies for national security. The genesis of that mission and of DARPA itself dates to the launch of Sputnik in 1957, and a commitment by the United States that, from that time forward, it would be the initiator and not the victim of strategic technological surprises. Working with innovators inside and outside of government, DARPA has repeatedly delivered on that mission, transforming revolutionary concepts and even seeming impossibilities into practical capabilities. The ultimate results have included not only game-changing military capabilities such as precision weapons and stealth technology, but also such icons of modern civilian society such as the Internet, automated voice recognition and language translation, and Global Positioning System receivers small enough to embed in myriad consumer devices."

Agency staff note: "DARPA explicitly reaches for transformational change instead of incremental advances. But it does not perform its engineering alchemy in isolation. It works within an innovation ecosystem that includes academic, corporate and governmental partners, with a constant focus on the nation's military services, which work with DARPA to create new strategic opportunities and novel tactical options. For decades, this vibrant, interlocking ecosystem of diverse collaborators has proven to be a nurturing environment for the intense creativity that DARPA is designed to cultivate.

"DARPA comprises approximately 220 government employees in six technical offices, including nearly 100 program managers, who together oversee about 250 research and development programs.

"DARPA goes to great lengths to identify, recruit and support excellent program managers—extraordinary individuals who are at the top of their fields and are hungry for the opportunity to push the limits of their disciplines. These leaders, who are at the very heart of DARPA's history of success, come from academia, industry and government agencies for limited stints, generally three to five years. That deadline fuels the signature DARPA urgency to achieve success in less time than might be considered reasonable in a conventional setting."

DARPA stands for the Defense Advanced Research Projects Agency. It is in charge of research and development of advanced technologies for the U.S. military.

DARPA adds: "Program managers address challenges broadly, spanning the spectrum from deep science to systems to capabilities, but ultimately they are driven by the desire to make a difference. They define their programs, set milestones, meet with their performers and assiduously track progress. But they are also constantly probing for the next big thing in their fields, communicating with leaders in the scientific and engineering community to identify new challenges and potential solutions.

"Program managers report to DARPA's office directors and their deputies, who are responsible for charting their offices' technical directions, hiring program managers and overseeing program execution. The technical staff is also supported by experts in security, legal and contracting issues, finance, human resources and communications. These are the people who make it possible for program managers to achieve big things during their relatively short tenures. At the Agency level, the DARPA Director and Deputy Director approve each new program and review ongoing programs, while setting Agency-wide priorities and ensuring a balanced investment portfolio.

"DARPA benefits greatly from special statutory hiring authorities and alternative contracting vehicles that allow the Agency to take quick advantage of opportunities to advance its mission. These legislated capabilities have helped DARPA continue to execute its mission effectively."

There is, however, a dark and disturbing side to DARPA and its work—not to mention its links to what we might call a New World Order, as we shall now see.

In 2010, Andrew Wozny of the *Canadian Internet Examiner* said: "No big surprise that DARPA, the military's blue-sky research arm, is the agency behind the lofty five-year program called Insight. The agency's goal is to replace 'largely manual exploitation and . . . chat-based operator interactions' with a system that mines different inputs, including drone footage and on-the-ground intelligence, and quickly stitches together the data to identify potential threats."

Wozny added: "The Pentagon's been investing in super-powered surveillance for years now, and DARPA wants Insight to capitalize on the rapid growth in the recon field.... And ongoing DARPA projects might be rolled into the Insight system too. The agency's solicitation cites a handful, including the recently launched PerSEAS, a program to design complex algorithms that can somehow spot threats based on little more than 'weak evidence.' It's the Information Awareness Office by another name. The Information Awareness Office was established by the Defense Advanced Research Projects Agency in January 2002 to bring together several DARPA projects focused on applying surveillance and information technology to track and monitor terrorists and other asymmetric threats to national security, by achieving Total Information Awareness....

"This would be achieved by creating *enormous computer databases* to gather and store the personal information of everyone in the United States, including personal emails, social network analysis, credit card records, phone calls, medical records, and numerous other sources, without any requirement for a search warrant.... Their logo says it all: the New World

An inventor and businesswoman, Regina Dugan was the first woman to serve as director of DARPA, which she managed from 2009 to 2012.

Order's pyramid and all-seeing Eye. It is the total tracking, total control, and total tyranny program."

In 2013, the situation became even graver. That was when news surfaced that a former director of the Defense Advanced Research Projects Agency (DARPA), Regina Dugan, was championing something that definitely raised both eyebrows and anxiety levels. It was what was described as nothing less than "an edible authentication microchip," designed to "contain a minute chip that transmits an individual's personal data."

Each chip is able to read a unique signal, a process that, if fully embraced, will sound the death knell for online passwords, passports, driver's licenses, and just about all private data. Rather than being implanted into a person's body on a one-time basis, it is inserted into a pill—a pill that a person will be required to swallow every day and that will be "designed to move through the body at the normal process of digestion, and according to engineers working on the device, it can be taken every day for up to a month."

Rather incredibly, the device—which, at the time I write these words, has already been officially approved by the Food and Drug Administration—is fueled by a battery that uses nothing less than the acid created in the stomach of the individual. When the chip and the acid interact, the former emits a minute signal that can be read by mobile devices, thereby determining and verifying the identity of the person.

Dugan, the head of advanced technology at Motorola, revealed that the company was developing just such a device—"a microchip inside a pill that users would swallow daily in order obtain the 'superpower' of having their entire body act as a biological authentication system for cellphones, cars, doors and other devices."

In stark terms, the human body itself becomes, as Dugan worded it, "your authentication token."

Infowars, which followed the story closely, noted: "Privacy advocates will wince at the thought, especially given Dugan's former role as head of DARPA, the Pentagon agency that many see as being at the top of the pyramid when it comes to the Big Brother technocracy."

Infowars also considered that Dugan's claim that such a pill could be taken an astonishing thirty times a day, every day, for the rest of one's natural life, was "seemingly dubious." They also asked the question: "Would you swallow a Google microchip every day simply to access your cellphone?"

Some might actually do that and much more, too. Dugan, in a fashion that must surely send chills up and down the spine of sane people everywhere, revealed that Motorola was working to develop a "wearable tattoo" that could essentially read the human mind by "detecting the unvocalized words in their throat."

John Hewitt of *Extreme Tech* explained the science behind this: "It has been known for decades that when you speak to yourself in your inner voice, your brain still sends neural spike volleys to your vocal apparatus, in a similar fashion to when you actually speak aloud."

Dugan apparently already had a method in place that would ensure a sizeable number of young people would eagerly sign up for the "e-tattoo." If it was designed in a fashion that was cool and rebellious, teenagers would quickly be on board, "if only to piss off their parents."

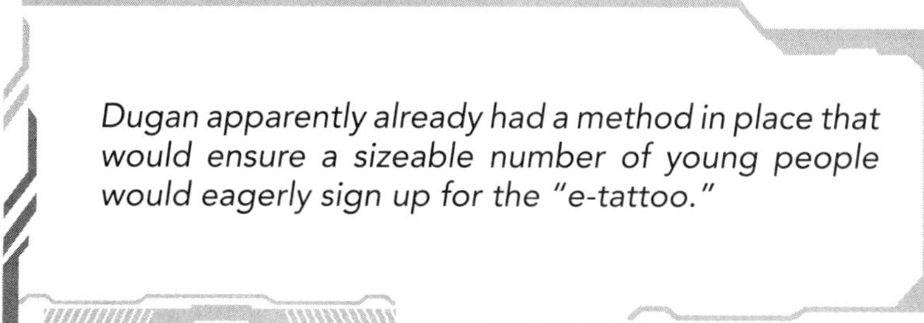
Dugan apparently already had a method in place that would ensure a sizeable number of young people would eagerly sign up for the "e-tattoo."

As *Infowars*'s Paul Joseph Watson noted: "The edible microchip and the wearable e-tattoo are prime examples of how transhumanism is being made 'trendy' in an effort to convince the next generation to completely sacrifice whatever privacy they have left in the name of faux rebellion (which is actually cultural conformism) and convenience."

Thankfully, some have warned of the growing threat. "The technology exists to create a totalitarian New World Order," noted Rauni-Leena Luukanen-Kilde, M.D., a former Finnish chief medical officer. She added: "Covert neurological communication systems are in place to counteract independent thinking and to control social and political activity on behalf of self-serving private and military interests. When our brain functions are already connected to supercomputers by means of radio implants and microchips, it will be too late for protest. This threat can be defeated only by educating the public, using available literature on biotelemetry and information exchanged at international congresses."

Greg Szymanski, who has followed the micro-chipping scandal, asks: "Are you ready for a total elimination of privacy and a robotizing of mankind, as well as an invasion of every thought going through your head? Are you prepared to live in a world in which every newborn baby is micro-chipped? And finally, are you ready to have your every move tracked, recorded and placed in Big Brother's data bank?"

When Artificial Intelligence Begins to Flex Its Muscles

NASA asks an important question: "Could the same computer algorithms that teach autonomous cars to drive safely help identify nearby asteroids or discover life in the universe? NASA scientists are trying to figure that out by partnering with pioneers in artificial intelligence (AI)—companies such as Intel, IBM and Google—to apply advanced computer algorithms to problems in space science." As for what, exactly, AI is, IBM defines it as follows: "It is the science and engineering of making intelligent machines, especially intelligent computer programs. It is related to the similar task of using computers to understand human intelligence, but AI does not have to confine itself to methods that are biologically observable."

On top of that, there are the words of the U.S. government's National Institute of Standards and Technology: "Artificial Intelligence (AI) is rapidly transforming our world. Remarkable surges in AI capabilities have led to a number of innovations including autonomous vehicles and connected Internet of Things devices in our homes. AI is even contributing to the development of a brain-controlled robotic arm that can help a paralyzed person feel again through complex direct human-brain interfaces. These new AI-enabled systems are revolutionizing everything from commerce and healthcare to transportation and cybersecurity.

"AI has the potential to impact nearly all aspects of our society, including our economy, but the development and use of the new technologies it brings are not without technical challenges and risks. AI must be developed in a trustworthy manner to ensure reliability, safety and accuracy."

Now, back to NASA: "Machine learning is a type of AI. It describes the most widely used algorithms and other tools that allow computers to learn from data in order to make predictions and categorize objects much faster and more accurately than a

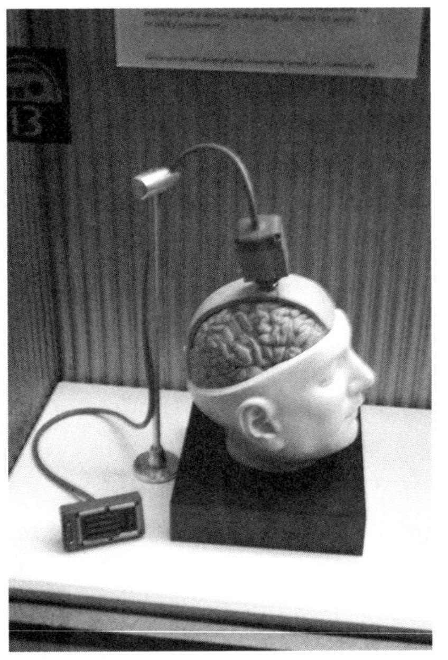

A computer-brain interface system called BrainGate was first developed by a U.S. company called Cyberkinetics and is now owned by BrainGate, Co. With it, patients who have lost limbs can control robotic ones with their thoughts.

human being can. Consequently, machine learning is widely used to help technology companies recognize faces in photos or predict what movies people would enjoy. But some scientists see applications far beyond Earth.

"Giada Arney, an astrobiologist at NASA's Goddard Space Flight Center in Greenbelt, Maryland, hopes machine learning can help her and her colleagues find a needle of life in a haystack of data that will be collected by future telescopes and observatories such as NASA's James Webb Space Telescope.

"'These technologies are very important, especially for big data sets and especially in the exoplanet field,' Arney says. 'Because the data we're going to get from future observations is going to be sparse and noisy. It's going to be really hard to understand. So using these kinds of tools has so much potential to help us.'

"To help scientists like Arney build cutting-edge research tools, NASA's Frontier Development Lab, or FDL, brings together technology and space innovators for eight weeks every summer to brainstorm and develop computer code. The four-year-old program is a partnership between the SETI Institute and NASA's Ames Research Center, both based in Silicon Valley where startup-hatching incubators that bring talented people together to accelerate the development of breakthrough technologies are abundant.

"In NASA's version, FDL pairs science and computer engineering early career doctoral students with experts from the space agency, academia, and some of the world's biggest technology companies. Partner companies contribute various combinations of hardware, algorithms, super-computer resources, funding, facilities and subject-matter experts. All of the AI techniques developed at FDL will be publicly available, with some already helping identify asteroids, find planets, and predict extreme solar radiation events.

"FDL feels like some really good musicians with different instruments getting together for a jam session in the garage, finding something really cool, and saying, 'Hey we've got a band here,' says Shawn Domagal-Goldman, a NASA Goddard astrobiologist who, together with Arney, mentored an FDL team in 2018. Their team developed a machine learning technique for scientists who aim to study the atmospheres of exoplanets, or planets beyond our solar system."

NASA's Jet Propulsion Lab AI Office provides this: "The Artificial Intelligence group performs basic research in the areas of Artificial Intelligence Planning and Scheduling, with applications to science analysis, spacecraft operations, mission analysis, deep space network operations, and space transportation systems. The Artificial Intelligence Group is organized administratively into two groups: Artificial Intelligence, Integrated Planning and Execution and Artificial Intelligence, Observation Planning and Analysis."

Furthermore, research into AI is now growing and expanding. Nature, for example, reveals the following: "China not only has the world's largest population . . . it also wants to lead the world when it comes to artificial intelligence (AI)."

Furthermore, research into AI is now growing and expanding. *Nature*, for example, reveals the following: "China not only has the world's largest population and looks set to become the largest economy—it also wants to lead the world when it comes to artificial intelligence (AI). In 2017, the Communist Party of China set 2030 as the deadline for this ambitious AI goal, and, to get there, it laid out a bevy of milestones to reach by 2020. These include making significant contributions to fundamental research, being a favored destination for the world's brightest talents, and having an AI industry that rivals global leaders in the field.

"As this first deadline approaches, researchers note impressive leaps in the quality of China's AI research. They also predict a shift in the nation's ability to retain homegrown talent. That is partly because the government has implemented some successful retainment programmes and partly because worsening diplomatic and trade relations mean that the United States—its main rival when it comes to most things, including AI—has become a less-attractive destination."

"In 2017, Russian President Vladimir Putin declared that whichever country becomes the leader in artificial intelligence (AI) 'will become the ruler of the world.'" Those were bold words, noted by Stephanie Petrella, Chris Miller, and Benjamin Cooper in the journal *Orbis* in 2021. The authors continued: "Yet Russia lags competitors like China and the United States substantially in AI capabilities. What is Russia's strategy for boosting development of AI technologies, and what role do groups within the Russian elite play in shaping this strategy?

"Russia's AI development strategy is unique in that it is led not by the government, nor by the private sector, but by state-owned firms. The government's distrust of Russia's largest tech firm, Yandex, has sidelined the company from national AI planning.

Russia's president, Vladimir Putin, is determined to make his nation the leader in AI technology because he believes it holds the key to dominating the world.

Meanwhile, Russia's defense conglomerate Rostec publicly appears to focus less on artificial intelligence than on other high-tech priorities. As a result, Russia's AI development has been left to a state-owned bank, Sberbank, which has taken the lead in devising plans for government-backed investment in AI."

From the appropriately titled *Future of Life*, consider these words: "In October 2019, the Office of the President of the Russian Federation released a national AI strategy, Decree of the President of the Russian Federation on the Development of Artificial Intelligence in the Russian Federation.... The strategy includes a long list of goals and primary objectives for the development of AI, including some to be achieved by the year 2024, and some by the year 2030. These goals include improving the availability and quality of data, increasing the availability of hardware, and creating appropriate standards and a regulatory system that guarantees public safety and stimulates the development of AI technologies."

Finally, DARPA has long-term plans when it comes to artificial intelligence. I'll share the agency's description with you, and without interruption, so that you'll see the sheer scale of the project: "The advance of technology has evolved the roles of humans and machines in conflict from direct confrontations between humans to engagements mediated by machines. Originally, humans engaged in primitive forms of combat. With the advent of the industrial era, however, humans recognized that machines could greatly enhance their warfighting capabilities. Networks then enabled teleoperation, which eventually proved vulnerable to electronic attack and subject to constraint due to long signal propagation distances and times. The next stage in warfare will involve more capable autonomous systems, but before we can allow such machines to supplement human warfighters, they must achieve far greater levels of intelligence.

"Traditionally, we have designed machines to handle well-defined, high-volume or high-speed tasks, freeing humans to focus on problems of ever-increasing complexity. In the 1950s and 1960s, early computers were automating tedious or laborious tasks. It was during this era that scientists realized it was possible to simulate human intelligence and the field of artificial intelligence (AI) was born. AI would be the means for enabling computers to solve problems and perform functions that would ordinarily require a human intellect.

"Early work in AI emphasized handcrafted knowledge, and computer scientists constructed so-called expert systems that captured the specialized knowledge of experts in rules that the system could then apply to situations of interest. Such 'first wave' AI technologies were quite successful—tax preparation software is a good example of an expert system—but the need to handcraft rules is costly and time-consuming and therefore limits the applicability of rules-based AI.

"The past few years have seen an explosion of interest in a sub-field of AI dubbed machine learning that applies statistical and probabilistic methods to large data sets to create generalized representations that can be applied to future samples. Foremost among these approaches are deep learning (artificial) neural networks that can be trained to perform a variety of classification and prediction tasks when adequate historical data is available. Therein lies the rub, however, as the task of collecting, labelling, and vetting data on which to train such 'second wave' AI techniques is prohibitively costly and time-consuming.

"DARPA envisions a future in which machines are more than just tools that execute human-programmed rules or generalize from human-curated data sets. Rather, the machines DARPA envisions will function more as colleagues than as tools. Towards this end, DARPA research and development in human-machine symbiosis sets a goal to partner with machines. Enabling computing systems in this manner is of critical importance because sensor, information, and communication systems generate data at rates beyond which humans can assimilate, understand, and act. Incorporating these technologies in military systems that collaborate with warfighters will facilitate better decisions in complex, time-critical, battlefield environments; enable a shared understanding of massive, incomplete, and contradictory information; and empower unmanned systems to perform critical missions safely and with high degrees of autonomy. DARPA is focusing its investments on a third wave of AI that brings forth machines that understand and reason in context."

Chipped Forever?

If you still doubt the possibility that we are going down a robotic pathway, one that will turn us into something very different from what we are right now, consider the next part of this story. Back in 2012, a highly controversial statement was made by one of the leading lights in science fiction. It came from Elizabeth Moon, who has a history degree from Rice University and a biology degree from the University of Texas at Austin. Her books include *The Serrano Connection*; *Engaging the Enemy*; and *Command Decision*. In 2012, Moon made a controversial statement that was highlighted by the BBC: "I would insist on every individual having a unique ID permanently attached—a barcode if you will—an implanted chip to provide an easy, fast, inexpensive way to identify individuals."

Microchips have proved very handy when it comes to the likes of tracking dogs that have gotten lost or stolen. The era of the microchip has ensured that what might otherwise have been sad and tragic affairs are far less common than they once were: time and again we hear of devoted owners reunited with their much-loved pets. That is, of course, a very good thing, but would it be a good thing for every single person on the planet to be implanted with a microchip? Some might say that having one's children chipped, in the event they get lost or, in a worst-case scenario, kidnapped, would be a very good thing. If that's all the chip was to be used for, then, yes, more than a few parents would likely sign up—maybe even the majority of the population.

The problem, however, is that implanted chip technology is wide open to abuse. Using the technology to find your lost pet, or a child who has gone missing, is almost certainly acceptable to many people, but what happens if government agencies wish to use the technology not just because you and your loved ones may be in peril but because they just want to know where you are on a 24/7 basis? I'll tell you what happens: freedom and personal privacy go right out the door. For your entire life. No wonder the issue is one that polarizes people into those who are pro-chip and those who are anti-chip.

It may seem like sci-fi, but it's not. Microchips work very well on animals. They work just as well when implanted in people. And day by day, the technology is getting ever more advanced. The *New York Daily News* asked an important question: "The proposal isn't too far-fetched—it is already technically possible to 'barcode' a human—but does it violate our rights to privacy?" Yes, it was an important question, but surely, anyone with even half a brain would realize that being chipped and watched on an unending basis is a violation of our personal rights.

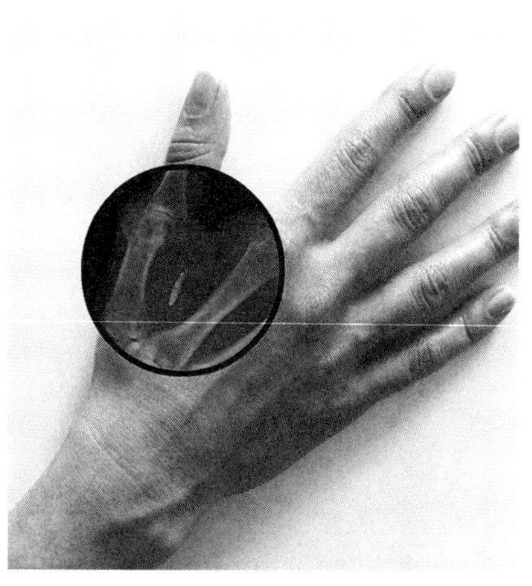

Chip implants are already being inserted subcutaneously into people, often in the hand. An RFID chip is used to identify you with a quick scan. The chip can contain a great deal of information about you, such as your location and medical history.

Far more forthright than the *New York Daily News* are the words of the American Civil Liberties Union: "To have a record of everywhere you go and everything you do would be a frightening thing. Once we let the government and businesses go down the road of nosing around in our lives, we're going to quickly lose all our privacy." It's good to see that at least one large and influential body is doing what it can to warn us all of the looming nightmare of endless tracking.

The *Inquirer* came right to the point when it said: "There is also the fear that this technology could be used by unscrupulous people or criminals, by competing corporations, or even by some agencies in the government, for illegal information gathering or surveillance, or for some immoral objectives."

Indeed, if your chip happens to contain, for example, your address, your phone number, your social security number, your email address, and your . . . well, you get the picture. If a sophisticated hacker could access your chip and obtain all of the above personal data and more, you could be in big trouble—all as a result of a small device that is buried just under the surface of your skin. And if that hacker happens to be in the employ of the National Security Agency, well, you really have no one to blame but yourself for becoming one of the sheeple.

But, here's the most disturbing aspect of all this: the process is already underway—and in earnest.

Certainly, the most visible example of human microchipping occurred in Florida in May of 2002. That was when the Jacobs family underwent the equivalent of taking the dogs to the local veterinary office.

Certainly, the most visible example of human microchipping occurred in Florida in May of 2002. That was when the Jacobs family underwent the equivalent of taking the dogs to the local veterinary office. Yes, each and every one of them got implanted: Jeffrey and Leslie and their son, Derek. The plan was to implant the family members with chips that would note all of their medical history, including any specific conditions they had. Again, some might say this is a good thing, particularly if someone has the likes of heart disease, diabetes, or epilepsy, where treatment might be needed immediately in an emergency.

The technology—known as the VeriChip—was the work of Applied Digital Solutions, also of Florida. They were delighted to have a family of guinea pigs to work with. So were the Jacobses' doctors, who gleefully expressed their hopes that it wouldn't be too long before having one's very own chip became the norm. Mom, Dad, and son drove to a clinic in Boca Raton, and before you could say "Gotcha!" all three were implanted and ready for tracking. Yes, it really was that easy and straightforward. Some commentators in the media suggested that this was a wonderful idea. After all, it would save the hassle of fumbling around in an emergency situation to see if a passed-out person had a diabetic bracelet—which was one particular scenario rolled out by the media. Applied Digital Solutions were over the moon, hoping that their product would soon be a worldwide one.

Other media outlets were far less enthused by all this. The BBC clearly had its mind on issues other than medical-related ones, as their words demonstrated: "The chips could also be used to contain personal information and even a global positioning device which could track a person's whereabouts, leading to fears the chip could be used for more sinister purposes."

The BBC was not alone. Two well-respected figures came forward to give their opinions on what they felt was the first step down a slippery slope toward complete, total surveillance. They were Dr. Kirstie Ball of the Open University, who is a lecturer in organization studies. The other was the managing editor of *Surveillance and Society*, Dr. David M. Wood. In 2006 they warned, with good reason, that before the end of the second decade of the twenty-first century, we could find "our almost every movement, purchase and communication" watched by "a complex network of interlinking surveillance technologies."

The words of Dr. Wood and Dr. Ball were not at all far off the mark. For example, more than ten years after they made their statement, just about everyone owns a smartphone. It's incredibly easy for the likes of the National Security Agency to monitor your every move via your phone. Granted, it's not implanted in your skin, but to say that most people's iPhones are practically implanted in their pocket or in their hand would not be too far off the reality of the situation.

Richard Thomas, a former United Kingdom information commissioner, said that the chips Wood and Ball warned about might, at some point in the near future, "be used by companies who want to keep tabs on an employee's movements or by Governments who want a foolproof way of identifying their citizens—and storing information about them."

Offering a perfect example of what such a nightmare might bring, *The American Dream*'s Michael Snyder posed this thought-provoking question in May 2012 to his readers: "What would you do if someday the government made it mandatory for everyone to receive an implantable microchip for identification purposes? Would you take it?"

You might have no choice—that is, unless you fight against such a thing. Indeed, the U.S. military is already using technology for its troops. Snyder astutely sees where all of this may well be heading: "Once the government has microchips implanted in all of our soldiers, how long will it be before they want to put a microchip in all government employees for the sake of national security? Once the government has microchips in all government employees, how long will it be before they want to put a microchip in you?"

Interestingly enough, the very debates that roiled in the early 2000s, and which are being expanded upon now, were anticipated close to a quarter of a century before by David Icke, the author of such books as *The Biggest Secret*, *Children of the Matrix*, and *The Robots' Rebellion*. In 1994, Icke warned those who would listen: "The Brotherhood of the New World Order . . . want us bar-coded so we can be 'read' at supermarkets and banks, like a checkout assistant now reads a tin of beans. A man at IBM who invented the laser-bar reader for supermarkets has also developed a method of putting the same type of device under human skin in one billionth of a second. It is invisible to the naked eye and could carry all the information anyone needed to know about us. We could be permanently linked to a computer, and who is to say that signals could not be sent both ways?"

Icke was not done. In 1995, he had even more to say: "The game plan is known as the Great Work of Ages or the New World Order, and it seeks to introduce a world government to which all nations would be colonies; a world central bank and currency; a world army; and a microchipped population connected to a global computer. What is happening today is the culmination of the manipulation which has been unfolding for thousands of years."

A former British football player and sports broadcaster, David Icke is now a professional conspiracy theorist.

Many laughed at Icke at the time—and particularly so the media of the U.K. They aren't laughing so much now.

There will always be those who will fly the flag of "Well, it was good enough for my dog, so it's good enough for me." Others might note that such implants would be very useful in health-related emergencies. If used responsibly and for the reasons directly above, then many would likely embrace the technology. The question, though, is this: How exactly can we be sure that the technology will be used responsibly? The answer is: We can't. We know that our smartphones are wide open to surveillance, hacking, and tracking. So are our computers and laptops. Why would you expect the controllers would not jump at the chance to access everything contained on your personal chip? Of course, they would. And given the chance, they will. It's up to us to not give them that chance.

Cloning: Inevitable for Us?

Let me introduce something new to ponder when it comes to the human race changing into something akin to a robot. It's time to focus our attention on a company titled Valiant Venture Limited. It was officially created in the second month of 1997. It oversaw another company, which had a far more intriguing name: Clonaid. That's right: we're talking about clones and what might accurately be termed as "grown people." Impressive amounts of financial offerings to Clonaid ensured that the organization was able to begin its research—in the Bahamas, no less. The plan was not just to create clones but to achieve nothing less than definitive immortality. In many respects, this sounds very much like the biological robots that Philip Corso claimed were found just outside the city of Roswell, New Mexico, in early July 1947. Interestingly, there is a UFO/alien angle to all of this. The man who oversaw the Clonaid venture claimed to have met extraterrestrials. His name: Raël. (Yes, he did use just one name.)

Clonaid says of its history and work: "CLONAID™, the first human cloning company in the world, was founded in February 1997, by RAEL and a group of investors who created the Valiant Venture Ltd Corporation based in the Bahamas. In the first couple of years CLONAID™ has already received enormous media coverage. However, due to the pressure mounted on the Bahamas government by French journalists, Valiant Venture Ltd was cancelled as government representatives were thinking the laboratories would be established on the Bahamas Island. Meanwhile, the list of serious potential customers had grown to more than 250 people! Therefore, during the year 2000, Raël decided to hand over the CLONAID™ project to Dr. Brigitte Boisselier, a Raëlian bishop, in order for her to start working on actually cloning the first human being with a team of well-trained scientists. Dr. Boisselier has PhD degrees in physical and biomolecular chemistry. In her last job she was a marketing director for a large chemical company in France.

"In the summer of 2000, an American couple that wanted to help develop this technology in order for them to have a baby contacted Dr. Boisselier. They were the first major investors funding the equipment and the laboratory needed and CLONAID™'s first human cloning laboratory was set up in early 2001. In the summer of 2001, following several visits from U.S. government representatives in our facilities, CLONAID™ decided to pursue its human cloning project in another country where human cloning is legal."

A French chemist with two Ph.D.s, Raëlian bishop Dr. Brigitte Boisselier claimed she has created the first human clone.

Of this highly controversial aspect of the story, the BBC said: "A controversial company linked to a UFO sect says it has produced the world's first cloned human baby. However, the announcement has been viewed with deep skepticism by the scientific community at large—and no proof has so far been put forward. At a press conference, US-based company Clonaid claimed the birth of a healthy cloned baby girl, nicknamed Eve by scientists, who was born by Caesarean section on Thursday to a 31-year-old US mother. The location of the alleged birth has been kept secret. The DNA to be cloned was taken from the mother's skin cell, Clonaid said.

"The scientist leading Clonaid's efforts, Dr Brigitte Boisselier, said she was 'celebrating a scientific success.' However a White House spokesman said that US President George W Bush had found the news 'deeply troubling,' adding that the news underscored the need for legislation to ban all human cloning in the US. And Fred Eckhard, a spokesman for United Nations Secretary-General Kofi Annan, said that in the absence of any scientific proof 'we can't automatically accept it as a fact. No-one should expect the secretary-general to send flowers,' he said."

With that said, let's look a bit closer at Raël. As is said: "On December 13, 1973, in the heart of the Puy de Lassolas crater, near Clermond-Ferrand in the center of France, a journalist, Claude Vorilhon, saw a metallic looking engine about 7 meters in diameter in the shape of a flattened bell descend from the sky. It resembled no existing terrestrial technology.

"Astounded, he saw the engine stop and a trap door open. A human being of small size (1.2 meters) descended and approached him. Reassured by the pacifist attitude of the visitor, Claude Vorilhon wanted to communicate, and questioned him in French, 'Where do you come from?'

"He heard the small being answer, 'From very far, from another planet. I have come to meet you, you Claude Vorilhon. I have many things to tell you, and I have chosen you for a difficult mission. You are going to transmit to humans what I am going to tell you, and according to their reactions we will see if we can officially show ourselves to them. I know that you have recently read the Bible. Come into my machine. We will be more comfortable to talk.'

"The extra-terrestrial was about four feet in height, had long dark hair, almond-shaped eyes, olive skin, and exuded harmony and humor. He gave Claude Vorilhon the name Rael, then told him, 'We were the ones who made all life on earth. You mistook us for gods. We were at the origin of your main religions. Now that you are mature enough to understand this, we would like to enter official contact through an embassy.'

"The messages dictated to Raël explain how life on Earth is not the result of random evolution, nor the work of a supernatural 'God.' It is a deliberate creation, using DNA, by a scientifically advanced people who made human beings literally in their image, what one can call 'scientific creationism.' On Wednesday, May 2, 2001, at the U.S. Senate, Subcommittee on Science, Technology, and Space, Committee on Commerce, Science, and Transportation, Washington, D.C., a lengthy statement was made. In part, it stated: 'Senator Brownback. The Committee will come to order. Today, we will be holding a hearing on the vital issue of human cloning. As our country debates the issue of human cloning, as well as those issues which surround it, I think it is helpful to engage in a similar dialog in the Senate. The issue of human cloning forces us to debate first principles—most particularly, the meaning of human life, and whether that life is a person or a piece of property. The importance of the issue of human cloning simply cannot be underestimated.

"'It is an issue that touches on our humanity in a way few issues have, and it does so at a time when we have the unique ability to resolve the issue properly, not only for our own country, but also to lead the way for the world. I recently introduced legislation to ban human cloning, Senate bill S. 790, which has been referred to the Senate Judiciary Committee. I believe there is a deep concern in America, and the world in general, with the use of this technology for the purposes of creating humans. In fact, according to a recent *Time*/CNN poll, 90 percent of Americans thought that it was a bad idea to clone human beings. I believe, along with Congressman Weldon, and many Americans share this belief, that efforts to create human beings by cloning mark a new and decisive step toward turning human reproduction into a manufacturing process in which children are made in laboratories to preordained specifications.

"'Creating cloned live-born children begins by creating cloned human embryos, a process which some also propose as a way to create embryos for research or as sources of cells and tissues for possible treatment of other humans. The prospect of creating new human life solely to be exploited and destroyed in this way has been condemned on moral grounds by many as displaying a profound disrespect for life. Furthermore, recent scientific advances indicate that there are fruitful and morally unproblematic alternatives to this approach. There is no need for this technology to ever be used with humans, whether for reproductive purposes or for destructive research purposes.'"

U.S. representative David Weldon of Florida, who served from 1995 to 2009, was one of many politicians who had severe reservations about the wisdom of cloning human beings.

There was more to come, much of it of a warning for those who were not in support of cloning humans, such as this statement from Rep. David Weldon of Florida: "Human cloning is the asexual reproduction of an organism which is genetically virtually identical to an existing or previously existing human being, performed by somatic cell nuclear transfer technology. It took 277 attempts to produce Dolly, and some estimate that producing a human child could take 1,000 attempts. Of cloned cows, sheep, goats, pigs, and mice, 95 to 97 percent of these efforts still end in failure. The attempt to clone humans does not account for the scientific problems which occur with animal cloning and, therefore, any such attempts will result in high failure rates.

"Most scientists agree that human cloning poses a serious risk of producing children who are stillborn, unhealthy, severely malformed or disabled, and almost universal opinion is that such attempts are thoroughly unethical. Additional problems with human cloning include the potential for mutation, transmission of mitochondrial diseases, and the negative effects from the aging genetic material.

"Abnormal clone development likely results from faulty DNA reprogramming, leading to abnormal gene expression of any of the 30,000 genes needed. Prenatal screening methods to detect chromosomal or genetic abnormalities in a fetus cannot detect these reprogramming errors, and no future methods exist for detecting reprogramming errors. If there would be undetectable genetic abnormalities in a developing human clone, then there may also be genetic defects in any tissues or cells derived from human clones.

"Cloning human beings is utilitarian in nature. Efforts to create human beings by cloning mark a new and decisive step toward turning human reproduction into a manufacturing process in which children are made in laboratories to preordained specifications and, potentially, in multiple copies. While people have indicated a desire to be cloned, almost no one has claimed that they would want to be a clone. Cloning could easily be used to reproduce persons without their consent. Because it is an asexual form of reproduction, cloning confounds the meaning of father and mother. Human cloning confuses the identity and kinship relations of any cloned child.

"This threatens to weaken existing notions regarding who bears which parental duties and responsibilities for children. Efforts to create human beings by cloning mark a new and decisive step toward turning human reproduction into a manufacturing process in which children are made in laboratories to preordained specifications and, potentially, in multiple copies. Some have stated that they want to clone themselves, but none have stated that they would want to be a human clone. Effects on the genetic characteristics of a population, have the potential to be used in a eugenic or discriminatory fashion. These practices are completely inconsistent with the ethical norms of medical practice. Moreover, the prospect of creating new human life solely to be exploited and destroyed for research purposes has been condemned on moral grounds by many, including supporters of a right to abortion, as displaying a profound disrespect for life. The General Board of Church and Society of the United Methodist is opposed to any form of human cloning."

Clearly, few favored the issue of cloning.

Since there was a great deal of confusion and lack of data on all of this, the U.S. Food and Drug Administration (FDA) came forward with vital, important data for the public, the media, and the government. In part, the FDA said the following due to the fact that there were worries not just about people being cloned, but animals, too: "Despite science fiction books and movies, clones are born just like any other animal. The only difference is that clones don't require a sperm and egg to come together to make an embryo. Clone embryos are made by using a whole cell or cell nucleus from a donor animal and fusing it to an egg cell that's had its nucleus removed. That embryo is implanted into the uterus of a surrogate dam (a livestock term that breeders use to refer to the female parent of an animal) to grow just as if it came from embryo transfer or *in vitro* fertilization."

The FDA added: "A clone produces offspring by sexual reproduction just like any other animal. A farmer or breeder can use natural mating or any other assisted reproductive technology, such as artificial insemination or *in vitro* fertilization to breed clones, just as they do for other farm animals. The offspring are not clones, and are the same as any other sexually reproduced animals.

"Clones are born the same way as other newborn animals: as babies. No one really knows what causes aging in mammals, but most scientists think it has to do with a part of the chromosome called a telomere that functions as a kind of clock in the cell. Telomeres tend to be long at birth, and shorten as the animal ages. A study on Dolly (the famous sheep clone) showed that her telomeres were the shorter length of her (older) donor, even though Dolly was much younger. Studies of other clones have shown that telomeres in clones are shorter in some tissues in the body, and are age-appropriate in other tissues. Still other studies of clones show that telomeres are age-appropriate in all of the tissues. Despite the length of telomeres reported in different studies, most clones appear to be aging normally. In fact, the first cattle clones ever produced are alive, healthy, and are 10 years old as of January 2008."

That the U.S. government felt the need to explain the issues of cloning in both animals and people demonstrates just how widespread the concern—and the technology—has grown.

The Rise of the Cyborgs

To demonstrate just how close we are to becoming half-human and half-robot, there's the matter of cyborgs. But what, exactly, are cyborgs? At the National Library of Medicine/National Center for Biotechnology Information, there is this article: "New biomedical technologies make it possible to replace parts of the human body or to substitute its functions. Examples include artificial joints, eye lenses and arterial stents. Newer technologies use electronics and software, for example in brain-computer interfaces such as retinal implants and the exoskeleton MindWalker. Gradually we are creating cyborgs: hybrids of man and machine. This raises the question: are cyborgs still humans? It is argued that they are. First, because employing technology is a typically human characteristic. Second, because in western thought the human mind, and not the body, is considered to be the seat of personhood. However, it has been argued by phenomenological philosophers that the body is more than just an object but is also a subject, important for human identity. From this perspective, we can appreciate that a bionic body does not make one less human, but it does influence the experience of being human."

Much of this research into the field of cyborgs is directed to making it feasible for our astronauts to be able to reach Mars and other faraway worlds. Right now, our bodies are not ready to take such long flights—and never mind the return flights. Using cyborgs would make things much easier, but it would also take away some of our humanity.

Some scientists believe that if we are going to successfully colonize Mars and other worlds, having cyborgs available to do much of the work would help make this a reality.

On this undeniably inflammatory issue, Amy Shira Teitel writes at *Vice*: "We could modify men on Earth to prepare them for life on Mars. We could start selectively breeding men right now to have favorable generic mutations such as a need for less oxygen or higher starting bone density, increasing life spans such that a single crew could make significant headway in setting up a base on Mars. Once on Mars, our custom-made astronaut would adapt to his new life much more easily than the run-of-the-mill human. Some of the most outspoken proponents of manned space exploration are in favor of creating a race of post-biological human: a being that would take the essence of a man—his rational faculties, intelligence, and personality—and put it in the durable outer casing of a machine."

Bhavya Sukheja at *RepublicWorld* has something intriguing to say on this matter: "SpaceX CEO Elon Musk on February 1 revealed an ambitious plan to get humans on Mars by 2026, which is seven years before US space agency NASA aims to land astronauts on the Red Planet. While speaking on the audio-only social media app Clubhouse, Musk said that his goal was to establish a self-sustaining Martian civilisation. For the first time, he mentioned a timeline and said that he will get humans on the Red Planet in 'five and a half years.'

"Musk's deadline seems little ambitious as there is a long way to go. SpaceX is still working to finalize the prototypes, with a second high-altitude test flight due soon. Even NASA aims to get first humans on the

Red Planet in 2033. It is also worth mentioning that it currently takes at least six months to get to Mars, however, Musk believes that could be down to as little as a month, with flights operating every two years."

Space.com says: "In 1960, Manfred Clynes and Nathan Kline published an essay in *Astronautics* titled 'Cyborgs in Space.' Comparing man in space to a fish out of water, they noted that even if you could bring everything you need on your space explorations, 'the

Entrepreneur and founder of such companies as Tesla and SpaceX, Elon Musk has ambitions to lead humanity to colonize Mars.

bubble all too easily bursts.' However, if the human body were altered to adapt to the conditions of space, astronauts would be free to explore the universe without limitation. 'Solving the many technical problems involved in manned space flight by adapting man to his environment, rather than vice versa, will not only mark a significant step forward in man's scientific progress, but may well provide a new and larger dimension for man's spirit as well,' the authors write."

"Humans are built to thrive on Earth, but even a yearlong round-trip mission to Mars could pose major medical risks. If we want humans to colonize the solar system, we may have to fundamentally alter our biology and become cyborgs." Those are the words of Alasdair Wilkins in *Gizmodo*. Wilkins continues: "That's the argument put forward by Smithsonian

Being ill-adapted to their new habitat, the pioneer Martian explorers will have a more compelling incentive than those of us on Earth to redesign themselves.

National Air and Space Museum senior curator Roger Launius, who believes unmodified humans won't be able to survive the unforgiving off-world environment. Between cosmic radiation, low gravity levels, and the constant threat of running out of the most basic resources like water or oxygen, the challenge might be too great for humans as they are now, which raises the question of using tactical mechanical enhancements to create cyborg colonists."

Writers at the *Daily Mail*, too, have recognized the good and bad side of all this, noting the "ethical grounds" in turning us into cyborgs: "The space environment is inherently hostile for humans. Being ill-adapted to their new habitat, the pioneer Martian explorers will have a more compelling incentive than those of us on Earth to redesign themselves. They'll harness powerful genetic and cyborg technologies that will be developed in coming decades. These techniques will be, one hopes, regulated on Earth, on prudential and ethical grounds. We should wish them good luck in modifying their progeny to adapt to alien environments. It's these space-faring adventurers, not those of us comfortably adapted to life on Earth, who will spearhead the post-human era."

Consider the following from *Seeker*: "If you've been following some of the recent developments in the world of medicine, you know that humans today can use their thoughts to manipulate machines, artificial limbs which interact with their nervous systems, and even use voice synthesizers to quite literally speak their mind if their ability to talk has been taken away from them. Much of this technology is still experimental, but as it's proving to be effective and safe, we can use it to re-engineer the bodies of astronauts to handle the rigors of space in other worlds. Basically, imagine space being explored not by people in space suits, but by cyborgs who were specially modified to cope with just about everything alien environments can throw at them."

Mind Control Madness

There's yet another way in which we may become far more like robots than humans. It's the field of mind control. Is it feasible that we, the human race, could be controlled to the extent that one day we might be relegated to nothing less than robot-like slaves for the worldwide elite? We saw the beginnings in the first page of the previous chapter, related to cyborgs. You'll see it even more in this chapter. Take a look now as to how such a nightmarish situation might become all too real. Within the annals of research into conspiracy theories, there is perhaps no more emotive term than that of "mind control." Indeed, mention those two words to anyone who is even remotely aware of the term and it will invariably and inevitably (and wholly justifiably, too) provoke imagery and comments pertaining to political assassinations, dark and disturbing CIA chicanery, sexual slavery, secret government projects, alien abductions, and subliminal advertising on the part of the world's media and advertising agencies. Yes, the specter of mind control is one that has firmly worked its ominous way into numerous facets of modern-day society. And it has been doing so for years. To say that, today, mind-control technology can turn us into robot-like, emotionless automatons is not an exaggeration.

"I can hypnotize a man, without his knowledge or consent, into committing treason against the United States," asserted Dr. George Estabrooks, chairman of the Department of Psychology at Colgate University, way back in 1942, before a select group of personnel attached to the United States' War Department. Estabrooks added: "Two hundred trained foreign operators, working in the United States, could develop a uniquely dangerous army of hypnotically controlled Sixth Columnists."

Estabrooks's pièce de résistance, however, was to capitalize on an ingenious plan that had been postulated as far back as World War I. As he explained: "During World War One, a leading psychologist made a startling proposal to the navy. He offered to take a submarine steered by a captured U-boat captain, placed under his hypnotic control, through enemy mine fields to attack the German fleet. Washington nixed the stratagem as too risky. First, because there was no disguised method by which the captain's mind could be outflanked. Second, because today's technique of 'day-by-day breaking down of ethical conflicts' brainwashing was still unknown.

"The indirect approach to hypnotism would, I believe, change the navy's answer today. Personally, I am convinced that hypnosis is a bristling, dangerous armament which makes it doubly imperative to avoid the war of tomorrow."

A perfect example of the way in which the will of a person could be completely controlled was amply and graphically spelled out in an article that Estabrooks wrote in April 1971 for the now-defunct publication *Science Digest*. Titled "Hypnosis Comes of Age," it stated the following:

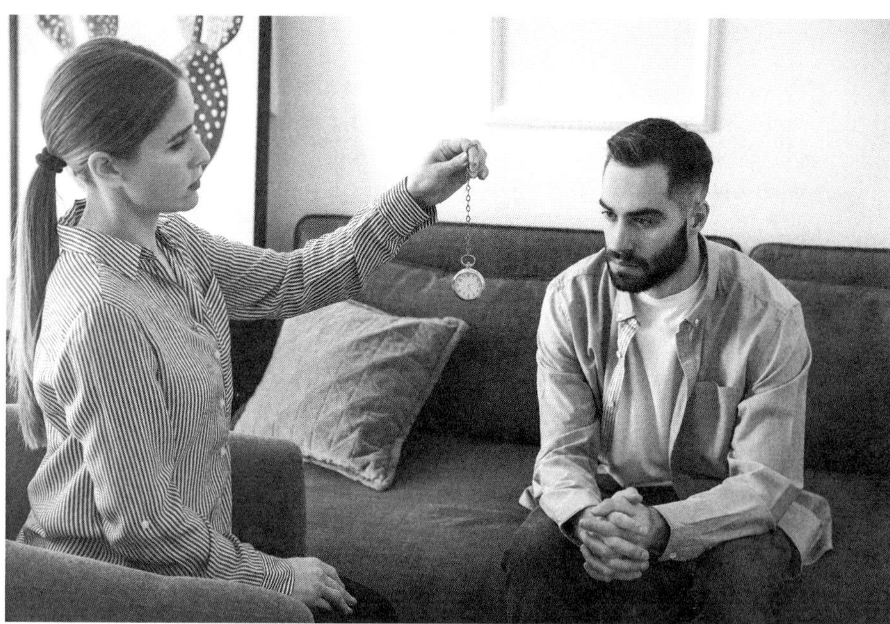

The military has long looked into the idea of hypnotizing people to become easily controlled agents, and that practice might be ongoing in certain government agencies.

"Communication in war is always a headache. Codes can be broken. A professional spy may or may not stay bought. Your own man may have unquestionable loyalty, but his judgment is always open to question.

"The 'hypnotic courier,' on the other hand, provides a unique solution. I was involved in preparing many subjects for this work during World War II. One successful case involved an Army Service Corps Captain whom we'll call George Smith.

"Captain Smith had undergone months of training. He was an excellent subject but did not realize it. I had removed from him, by post-hypnotic suggestion, all recollection of ever having been hypnotized.

"First I had the Service Corps call the captain to Washington and tell him they needed a report of the mechanical equipment of Division X headquartered in Tokyo. Smith was ordered to leave by jet next morning, pick up the report and return at once. Consciously, that was all he knew, and it was the story he gave to his wife and friends.

"Then I put him under deep hypnosis, and gave him—orally—a vital message to be delivered directly on his arrival in Japan to a certain colonel—let's say his name was Brown—of military intelligence.

"Outside of myself, Colonel Brown was the only person who could hypnotize Captain Smith. This is 'locking.'

"I performed it by saying to the hypnotized Captain: 'Until further orders from me, only Colonel Brown and I can hypnotize you. We will use a signal phrase *the moon is clear*. Whenever you hear this phrase from Brown or myself you will pass instantly into deep hypnosis.'

"When Captain Smith re-awakened, he had no conscious memory or what happened in trance. All that he was aware of was that he must head for Tokyo to pick up a division report.

"On arrival there, Smith reported to Brown, who hypnotized him with the signal phrase. Under hypnosis, Smith delivered my message and received one to bring back. Awakened, he was given the division report and returned home by jet. There I hypnotized him once more with the signal phrase, and he spieled off Brown's answer that had been dutifully tucked away in his unconscious mind."

And with the early, groundbreaking work of George Estabrooks now concisely spelled out for one and all to read, digest, and muse upon, let me acquaint you with a concise history of the world of mind control, mind manipulation, and what could accurately be termed mind slavery.

Although the U.S. intelligence community, military, and government have undertaken countless official projects . . . pertaining to both mind control and mind manipulation, without doubt the most notorious of all was Project MKUltra. . . .

The picture is not a pretty one—not at all.

Although the U.S. intelligence community, military, and government have undertaken countless official projects (and off-the-record ones, too) pertaining to both mind control and mind manipulation, without doubt the most notorious of all was Project MKUltra: a clandestine operation that operated out of the CIA's Office of Scientific Intelligence and had its beginnings in the Cold War era of the early 1950s.

The date of the project's actual termination is a somewhat hazy one; however, it is known that it was definitely in operation as late as the latter part of the 1960s and, not surprisingly and regretfully, has since been replaced by far more controversial and deeply hidden projects.

To demonstrate the level of secrecy that surrounded Project MKUltra, even though it had kicked off at the dawn of the fifties, its existence was largely unknown outside the intelligence world until 1975, when the Church Committee and the Rockefeller Commission began making their own investigations of the CIA's mind control–related activities—in part to determine if (a) the CIA had engaged in illegal activity, (b) the personal rights of citizens had been violated, and (c) if the projects at issue had resulted in fatalities—which they most assuredly and unfortunately did.

Rather conveniently—and highly suspiciously—it was asserted at the height of the inquiries in 1975 that two years earlier, in 1973, CIA director Richard Helms had ordered the destruction of the agency's MKUltra files. Fortunately, this did not stop the Church Committee or the Rockefeller Commission—both of whom had the courage and tenacity to forge ahead with their investigations, relying on sworn testimony from players in MKUltra, where documentation was no longer available for scrutiny and evaluation.

The story that unfolded was both dark and disturbing, in equal degrees. Indeed, the scope of the project—and allied operations, too—was spelled out in an August 1977 document titled *The Senate MK-Ultra Hearings* that was prepared by the Senate Select Committee on Intelligence and the Committee on Human Resources, as a result of its probing into the secret world of the CIA.

As the document explained: "Research and development programs to find materials which could be used to alter human behavior were initiated in the late 1940s and early 1950s. These experimental programs originally included testing of drugs involving witting human subjects, and culminated in tests using unwitting, non-volunteer human subjects. These tests were designed to determine the potential effects of chemical or biological agents when used operationally against individuals unaware that they had received a drug."

The committee then turned its attention to the overwhelming secrecy that surrounded these early 1940s and 1950s projects: "The testing programs were considered highly sensitive by the intelligence agencies administering them. Few people, even within the agencies, knew of the programs and there is no evidence that either the Executive Branch or Congress were ever informed of them.

"The highly compartmented nature of these programs may be explained in part by an observation made by the CIA Inspector General that, 'the knowledge that the Agency is engaging in unethical and illicit activities would have serious repercussions in political and diplomatic circles and would be detrimental to the accomplishment of its missions.'"

The research and development programs, and particularly the covert testing programs, resulted in massive abridgments of the rights of American citizens, sometimes with tragic consequences.

The research and development programs, and particularly the covert testing programs, resulted in massive abridgments of the rights of American citizens, sometimes with tragic consequences. As prime evidence of this, the committee uncovered details on the deaths of two Americans that were firmly attributed to the programs at issue, while other participants in the testing programs were said to still be suffering from the residual effects of the tests as late as the mid-1970s.

And as the committee starkly noted: "While some controlled testing of these substances might be defended, the nature of the tests, their scale, and the fact that they were continued for years after the danger of surreptitious administration of LSD to unwitting individuals was known, demonstrate a fundamental disregard for the value of human life."

There was far more to come: The select committee's investigation of the testing and use of chemical and biological agents also raised serious questions about the adequacy of command-and-control procedures within the Central Intelligence Agency and military intelligence, and also about the nature of the relationships among the intelligence agencies, other governmental agencies, and private institutions and individuals that were also allied to the early mind-control studies.

For example, the committee was highly disturbed to learn that with respect to the mind-control and mind-manipulation projects, the CIA's normal administrative controls were controversially—and completely—waived for programs involving chemical and biological agents—supposedly to protect their security, but more likely to protect those CIA personnel who knew they were verging upon (if not outright surpassing) breaking the law.

But it is perhaps the following statement from the committee that demonstrates the level of controversy that surrounded—and still surrounds—the issue of mind control–based projects: "The decision to institute one of the Army's LSD field testing projects had been based, at least in part, on the finding that no long-term residual effects had ever resulted from the drug's administration. The

The CIA has tested LSD on unwitting Americans in an attempt to control their minds, according to some sources.

CIA's failure to inform the Army of a death which resulted from the surreptitious administration of LSD to unwitting Americans, may well have resulted in the institution of an unnecessary and potentially lethal program."

The committee added: "The development, testing, and use of chemical and biological agents by intelligence agencies raises serious questions about the relationship between the intelligence community and foreign governments, other agencies of the Federal Government, and other institutions and individuals.

"The questions raised range from the legitimacy of American complicity in actions abroad which violate American and foreign laws to the possible compromise of the integrity of public and private institutions used as cover by intelligence agencies."

While MKUltra was certainly the most infamous of all the CIA-initiated mind-control programs, it was very far from being an isolated one. Indeed, numerous sub-projects, post-projects and operations initiated by other agencies were brought to the committee's attention. One was Project Chatter, which the committee described thus:

"Project Chatter was a Navy program that began in the fall of 1947. Responding to reports of amazing results achieved by the Soviets in using truth drugs, the program focused on the identification and the testing of such drugs for use in interrogations and in the recruitment of agents. The research included laboratory experiments on animals and human subjects involving *Anabasis aphylla,* scopolamine, and mescaline in order to determine their speech-inducing qualities. Overseas experiments were conducted as part of the project. The project expanded substantially during the Korean War, and ended shortly after the war, in 1953."

Then there was Projects Bluebird and Artichoke. Again, the committee dug deep and uncovered some controversial and eye-opening data and testimony: "The earliest of the CIA's major programs involving the use of chemical and biological agents, Project Bluebird, was approved by the Director in 1950. Its objectives were: (a) discovering means of conditioning personnel to prevent unauthorized extraction of information from them by known means, (b) investigating the possibility of control of an individual by application of special interrogation techniques, (c) memory enhancement, and (d) establishing defensive means for preventing hostile control of Agency personnel."

The committee added with respect to Bluebird: "As a result of interrogations conducted overseas during the project, another goal was added—the evaluation of offensive uses of unconventional interrogation techniques, including hypnosis and drugs. In August 1951, the project was renamed Artichoke. Project Artichoke included in-house experiments on interrogation techniques, conducted 'under medical and security controls which would ensure that no damage was done to individuals who volunteer for the experiments. Overseas interrogations utilizing a combination of sodium pentothal and hypnosis after physical and psychiatric examinations of the subjects were also part of Artichoke.'"

Interestingly, the committee noted: "Information about Project Artichoke after the fall of 1953 is scarce. The CIA maintains that the project ended in 1956, but evidence suggests that Office of Security and Office of Medical Services use of 'special interrogation' techniques continued for several years thereafter."

MKNaomi was another major CIA program in this area. In 1967, the CIA summarized the purposes of MKNaomi thus: "(a) To provide for a covert support base to meet clandestine operational requirements. (b) To stockpile severely incapacitating and lethal materials for the specific use of TSD [Technical Services Division]. (c) To maintain in operational readiness special and unique items for the dissemination of biological and chemical materials. (d) To provide for the required surveillance, testing, upgrading, and evaluation of materials and items in order to assure absence of defects and complete predictability of results to be expected under operational conditions."

Under an agreement reached with the Army in 1952, the Special Operations Division (SOD) at Fort Detrick was to assist CIA in developing, testing, and maintaining biological agents and delivery systems....

Under an agreement reached with the Army in 1952, the Special Operations Division (SOD) at Fort Detrick was to assist CIA in developing, testing, and maintaining biological agents and delivery systems—some of which were directly related to mind-control experimentation. By this agreement, the CIA finally acquired the knowledge, skill, and facilities of the Army to develop biological weapons specifically suited for CIA use.

The committee also noted:

"SOD developed darts coated with biological agents and pills containing several different biological agents which could remain potent for weeks or months. SOD developed a special gun for firing darts coated with a chemical which could allow CIA agents to incapacitate a guard dog, enter an installation secretly, and return the dog to consciousness when leaving. SOD scientists were unable to develop a similar incapacitant for humans. SOD also physically transferred to CIA personnel biological agents in 'bulk' form, and delivery devices, including some containing biological agents."

In addition to the CIA's interest in using biological weapons and mind control against humans, it also asked SOD to study use of biological agents against crops and animals. In its 1967 memorandum, the CIA stated: "Three methods and systems for carrying out a covert attack against crops and causing severe crop loss have been developed and evaluated under field conditions. This was accomplished in anticipation of a requirement which was later developed but was subsequently scrubbed just prior to putting into action."

The committee concluded with respect to MKNaomi that the project was "terminated in 1970. On November 25, 1969, President Nixon renounced the use of any form of biological weapons that kill or incapacitate and ordered the disposal of existing stocks of bacteriological weapons. On February 14, 1970, the President clarified the extent of his earlier order and indicated that toxins—chemicals that are not living organisms but are produced by living organisms—were considered biological weapons subject to his previous directive and were to be destroyed. Although instructed to relinquish control of material held for the CIA by SOD, a CIA scientist acquired approximately 11 grams of shellfish toxin from SOD personnel at Fort Detrick which were stored in a little-used CIA laboratory where it went undetected for five years."

Recognizing, however, that when it came to mind control and manipulation, MKUltra was the one project that more than any other was worth pursuing as part of its efforts to determine the extent to which the CIA had bent and broken the law and flouted the rights of citizens, the committee had far more to say on the operation. Time and again the committee returned to Project MKUltra. This was not surprising, as it was, after all, the principal CIA program involving the research and development of chemical and biological agents, and was, in the words of the committee, "concerned with the research and development of chemical, biological, and radiological materials capable of employment in clandestine operations to control human behavior."

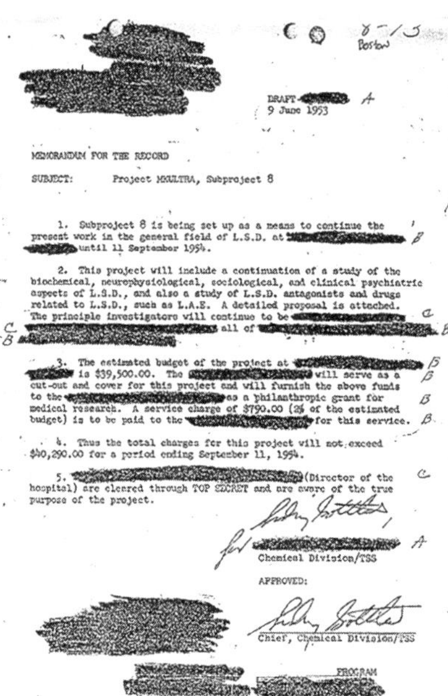

DRAFT-
9 June 1953

MEMORANDUM FOR THE RECORD

SUBJECT: Project MKULTRA, Subproject 8

1. Subproject 8 is being set up as a means to continue the present work in the general field of L.S.D. at until 11 September 1954.

2. This project will include a continuation of a study of the biochemical, neurophysiological, sociological, and clinical psychiatric aspects of L.S.D., and also a study of L.S.D. antagonists and drugs related to L.S.D., such as L.A.E. A detailed proposal is attached. The principle investigators will continue to be all of

3. The estimated budget of the project at is $39,500.00. The will serve as a cut-out and cover for this project and will furnish the above funds to the as a philanthropic grant for medical research. A service charge of $790.00 (2% of the estimated budget) is to be paid to the for this service.

4. Thus the total charges for this project will not exceed $40,290.00 for a period ending September 11, 1954.

5. (Director of the hospital) are cleared through TOP SECRET and are aware of the true purpose of the project.

Chemical Division/TSS

APPROVED:

Chief, Chemical Division/TSS

PROGRAM

A 1953 letter signed by CIA director Sidney Gottlieb authorizing LSD experiments.

The inspector general's survey of MKUltra, in 1963, noted the following reasons for the profound level of sensitivity that surrounded the program:

A. Research in the manipulation of human behavior is considered by many authorities in medicine and related fields to be professionally unethical, therefore the reputation of professional participants in the MKUltra program are on occasion in jeopardy.

B. Some MKUltra activities raise questions of legality implicit in the original charter.

C. A final phase of the testing of MKUltra products places the rights and interests of U.S. citizens in jeopardy.

D. Public disclosure of some aspects of MKUltra activity could induce serious adverse reaction in U.S. public opinion, as well as stimulate offensive and defensive action in this field on the part of foreign intelligence services.

Over the at least ten-year life of the program, many "additional avenues to the control of human behavior" were designated as being wholly appropriate for investigation under the MKUltra charter. These included "radiation, electroshock, various fields of psychology, psychiatry, sociology, and anthropology, graphology, harassment substances, and paramilitary devices and materials."

Needless to say, this was a grim list.

A 1955 MKUltra document provides a good example of the scope of the effort to understand the effects of mind-altering substances on human beings, and it lists those substances as follows. In the CIA's own words:

1. Substances which will promote illogical thinking and impulsiveness to the point where the recipient would be discredited in public.

2. Substances which increase the efficiency of mentation and perception.

3. Materials which will prevent or counteract the intoxicating effect of alcohol.

4. Materials which will promote the intoxicating effect of alcohol.

5. Materials which will produce the signs and symptoms of recognized diseases in a reversible way so that they may be used for malingering, etc.

6. Materials which will render the induction of hypnosis easier or otherwise enhance its usefulness.

7. Substances which will enhance the ability of individuals to withstand privation, torture and coercion during interrogation and so-called "brain-washing."

8. Materials and physical methods which will produce amnesia for events preceding and during their use.

9. Physical methods of producing shock and confusion over extended periods of time and capable of surreptitious use.

10. Substances which produce physical disablement such as paralysis of the legs, acute anemia, etc.

11. Substances which will produce "pure" euphoria with no subsequent let-down.

12. Substances which alter personality structure in such a way that the tendency of the recipient to become dependent upon another person is enhanced.

13. A material which will cause mental confusion of such a type that the individual under its influence will find it difficult to maintain a fabrication under questioning.

14. Substances which will lower the ambition and general working efficiency of men when administered in undetectable amounts.

15. Substances which promote weakness or distortion of the eyesight or hearing faculties, preferably without permanent effects.

16. A knockout pill which can surreptitiously be administered in drinks, food, cigarettes, as an aerosol, etc., which will be safe to use, provide a maximum of amnesia, and be suitable for use by agent types on an ad hoc basis.

17. A material which can be surreptitiously administered by the above routes and which in very small amounts will make it impossible for a man to perform any physical activity whatsoever.

In other words, when it came to mind manipulation, more than half a century ago, the CIA already had all bases covered.

A special procedure, designated MKDelta, was established to govern the use of MKUltra materials when specifically utilized in overseas operations. Such materials were used on a number of occasions. According to the committee: "Because MKUltra records were destroyed, it is impossible to reconstruct the operational use

The MKUltra plan for using drugs to manipulate and control people's minds dates back to the mid-1950s.

of MKUltra materials by the CIA overseas; it has been determined that the use of these materials abroad began in 1953, and possibly as early as 1950."

The committee expanded further: "Drugs were used primarily as an aid to interrogations, but MKUltra/MKDelta materials were also used for harassment, discrediting, or disabling purposes. According to an Inspector General Survey of the Technical Services Division of the CIA in 1957—an inspection which did not discover the MKUltra project involving the surreptitious administration of LSD to unwitting, non-volunteer subjects—the CIA had developed six drugs for operational use, and they had been used in six different operations on a total of thirty-three subjects. By 1963 the number of operations and subjects had increased substantially."

Aside from the CIA, the committee learned that the U.S. Army was up to its neck in mind control–related projects too. In its 1977 report, the committee wrote:

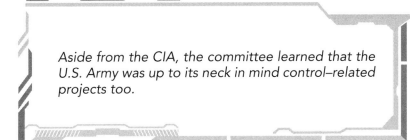

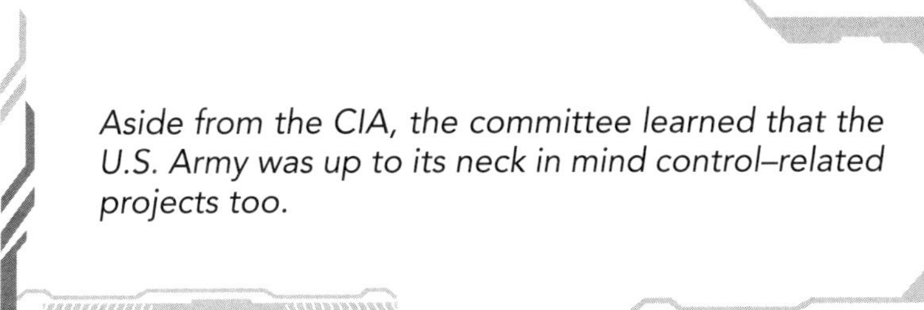
Aside from the CIA, the committee learned that the U.S. Army was up to its neck in mind control–related projects too.

"There were three major phases in the Army's testing of LSD. In the first, LSD was administered to more than 1,000 American soldiers who volunteered to be subjects in chemical warfare experiments. In the second phase, Material Testing Program EA 1729, 95 volunteers received LSD in clinical experiments designed to evaluate potential intelligence uses of the drug. In the third phase, Projects Third Chance and Derby Hat, 16 unwitting non-volunteer subjects were interrogated after receiving LSD as part of operational field tests."

But what of the post-MKUltra era: did the official world really cease its operations and destroy its files en masse, in 1973, as had been alleged? Probably not. In a 1977 interview, 14-year CIA veteran Victor Marchetti stated that the CIA's claim that MKUltra was abandoned was nothing more than a "cover story."

In conclusion, it comes down to this: one day, using post-MKUltra-syle technology, we just might find ourselves mind controlled—and in nothing less than slavish, robotic fashion.

Now let's go to the post-MKUltra era: a time when even more controversial work was done to turn people into, in essence, robots. In the 1950s and 1960s, a large number of experiments in behavior modification were conducted in the United States, and it is well known that electrical implants were inserted into the brains of animals and people. Later, when new techniques in influencing brain functions became a priority to military and intelligence services, secret experiments were conducted with such unwilling guinea pigs as inmates of prisons, mental patients, handicapped children, the elderly, and just about any and all groups considered expendable. It is an area of research that would have the control-obsessed New World Order practically hyperventilating.

The late Rauni-Leena Luukanen-Kilde, M.D., a former chief medical officer in Finland, has stated that mysterious brain implants, the size of one centimeter, began showing up on X-rays in the 1980s. In a few years implants were found the size of a grain of rice. Dr. Luukanen-Kilde stated that the implants were at first made of silicon, later of gallium arsenide. Today, such implants are small enough that it is nearly impossible to detect or remove them. They can easily be inserted into the neck or back during surgical operations, with or without the consent of the subject.

It has been stated that within a few years all Americans will be forced to receive a programmable biochip implant somewhere in their body—a key aspect of the New World Order agenda. The biochip is most likely to be implanted on the back of the right or left hand so it will be easy to scan at stores. The biochip will also be used as a universal type of identification card. A number will be assigned at birth and will follow that person throughout their life. Eventually, every newborn will be injected with a microchip, which will identify the person for the rest of his or her life.

Initially, people will be informed that the biochip will be used largely for purposes of identification. The reality is that the implant will be linked to a massive supercomputer system....

Initially, people will be informed that the biochip will be used largely for purposes of identification. The reality is that the implant will be linked to a massive supercomputer system that will make it possible for government agencies to maintain a surveillance of all citizens by ground sensors and satellites. Today's microchips operate by means of low-frequency radio waves that target them. With the help of satellites, the implanted person can be followed anywhere. Their brain functions may be remotely monitored by supercomputers and even altered through the changing of frequencies. Even worse, say the alarmists, once the surveillance system is in place, the biochips will be implemented to transform every man, woman, and child into a controlled slave, for these devices will make it possible for outside intelligences to influence a person's brain-cell conversations and talk directly with the individual's brain neurons. Through cybernetic biochip brain implants, people may be forced to think and to act exactly as government intelligence agencies have preprogrammed them to think and act.

The technology exists right now to create a New World Order served by the zombie-like masses. Secret government agencies could easily utilize covert neurological communication systems to subvert independent thinking and to control social and political activity.

The National Security Agency's electronic surveillance system can simultaneously follow the unique bioelectrical resonance brain frequency of millions of people. The NSA's Signals Intelligence (SIGINT) group can remotely monitor information from human brains by decoding the evoked potentials (3.50 hertz, 5 milliwatts) emitted by the brain. Electromagnetic frequency (EMF) brain stimulation signals can be sent to the brains of specific individuals, causing the desired effects to be experienced by the target.

A U.S. Navy research laboratory, funded by intelligence services, has achieved the incredible breakthrough of uniting living brain cells with microchips. When such a chip is injected into a person's brain, they instantly become a living vegetable and a subservient slave.

Experts have said that a micromillimeter microchip may be placed into the optical nerve of the eye and raw neuro-impulses from the brain that embody the experiences, smells, sights, and voice of the implanted subject. These neuro-impulses may be stored in a computer and may be projected back to the person's brain via the microchip to be re-experienced. A computer operator can send an electromagnetic message to the target's nervous system, thereby inducing hallucinations. Beyond all science-fiction scenarios, we could become a nation of New World Order–controlled robot-zombies.

A report titled "Biological Effects of Electromagnetic Radiation (Radiowaves and Microwaves) Eurasian Communist Countries," written in 1976 by Ronald L. Adams and Dr. R. A. Williams of the U.S. Army's Medical Intelligence and Information Agency, notes in part: "The Eurasian Communist countries are actively involved in evaluation of the biological significance of radiowaves and microwaves. Most of the research being conducted involves animals or in vitro evaluations, but active programs of a retrospective nature designed to elucidate the effects on humans are also being conducted."

Of deep concern to the United States was the incredible revelation that the Soviets had developed technology that allowed them to beam "messages" into the minds of targeted individuals.

Of deep concern to the United States was the incredible revelation that the Soviets had developed technology that allowed them to beam "messages" into the minds of targeted individuals. Rather notably, the Defense Intelligence Agency (DIA) and the U.S. Army concluded that such messages might direct a person to commit nothing less than suicide. Even if the person was not depressed, said Adams and Williams, the technology could be utilized to plunge them into sudden states of "irritability, agitation, tension, drowsiness, sleeplessness, depression, anxiety, forgetfulness, and lack of concentration."

The authors added: "Sounds and possibly even words which appear to be originating intracranially can be induced by signal modulation at very low average-power densities."

They concluded: "The Soviets will continue to investigate the nature of internal sound perception. Their research will include studies on perceptual distortion and other psycho-physiological effects. The results of these investigations could have military applications if the Soviets develop methods for disrupting or disturbing human behavior."

An extract from the Defense Intelligence Agency's "Controlled Offensive Behavior—USSR" document of 1972 might prove instructive. This one is particularly creepy, as it focuses on the possibility of secretly using ESP to hypnotize people and control them in—yes, you've got it—robotic, emotionless fashion. As you'll see, it's a subject that elements of U.S. intelligence were worried about during the Cold War. Maybe they still are. We're told the following: "According to [Sheila] Ostrander and [Lynn] Schroeder, the ability to put people to sleep and wake them up telepathically from a distance of a few yards to over a thousand miles became the most thoroughly tested and perfected contribution of the Soviets to international parapsychology.

"It is reported that the ability to control a person's consciousness with telepathy is being further studied and tested in laboratories in Leningrad and Moscow. The work was started in the early 1920s but was not publicized until the early 1960s. The work was begun by K.O. Kotkov, a psychologist from Kharkov University, in 1924. Kotkov could telepathically obliterate an experimental subject's consciousness from short distances or from the opposite side of town. The work was documented by [L. L.] Vasilev who conducted research of his own but could not reveal it under Stalin's regime. The reality of telepathic sleep-wake, backed by columns of data, might be the most astonishing part of Vasilev's experiments in mental suggestion."

Now let's see what else the United States learned: "Parapsychologists in Leningrad and Moscow are involved in the telepathic manipulation of consciousness, now recording successes with the EEG. Doctor V. Raikov is involved in this EEG research as well as E. Naumov. Naumov reports that mental telepathy woke up a hypnotized subject (by telepathy) six of eight times. Naumov remarked that as soon as the telepathic 'wake up' is sent, trance becomes less and less deep, full consciousness returning in twenty to thirty seconds. In the Leningrad laboratory of Doctor Paul Gulyaiev (Bekhterev Brain Institute), friends of subjects have been trained to put them to sleep telepathically."

The following question was asked within the Pentagon: "Why are the Soviets again hard at work on the telepathic control of consciousness?" The Defense Intelligence Agency responded: "Doctor I. Kogan, like Vasilev, is probably doing it for theoretical reasons; still trying mathematically to prove that an electromagnetic carrier of telepathy is possible. Why other scientists may be delving into control of consciousness by ESP is another question. During telepathic sleep is an individual simply dreaming his own private dreams or does someone else hold sway? The current Soviets have not divulged the psychological details about their telepathic manipulation of consciousness. Vasilev describes some revelations in his book but little else has been reported. Doctor Stefan Manczarski of Poland predicts that this new field of telepathy will open up new avenues for spreading propaganda. He feels that the electromagnetic theory is valid and believes, therefore, that telepathy can be amplified like radio waves. Telepathy would then become a subtle new modus for the 'influences' of the world. Doctor Manczerski's wave ideas are still very debatable, but what about telepathy someday becoming a tool for influencing people?"

The DIA added: "Hypnotizing someone telepathically probably comes over as a more eerie, mystifying, almost diabolical act in the US than it does in the Soviet Union. The US is really just becoming adjusted to some of the aspects of hypnotism. Since the turn of the century, the Soviets have been exploring and perfecting the various advantages that hypnotism provides. In the Soviet Union hypnotism is a common tool like X-rays, used in medicine, psychotherapy physiology, psychology, and experimental pedagogy. The Soviets have been reportedly working on the effects of drugs used in combination with psychic tests. Vasilev used mescarine in the early days and more recently M. S. Smirnov, of the Laboratory of Vision, Institute of Problems of Information Transmission of the USSR Academy of Science, has been obtaining psychic success with psilocybin. The tests that Vasilev had perfected may have a more interesting future in them than the developer had imagined. Manipulating someone else's consciousness with telepathy, guiding him in trance colorful uses are too easy to conjure. The ability to focus a mental whammy on an enemy through hypnotic telepathy has surely occurred to the Soviets."

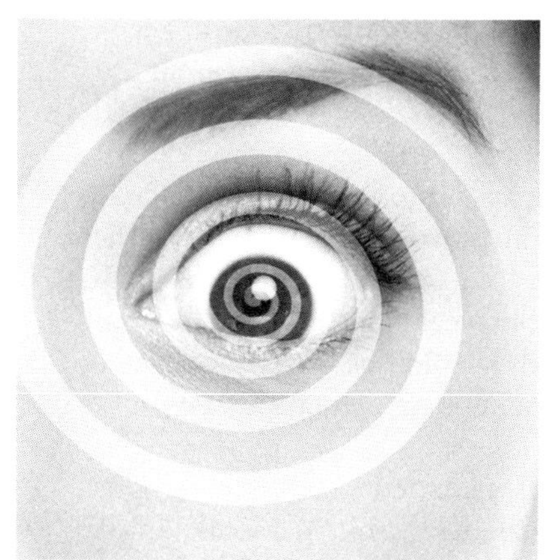

Besides drugs, telepathic control of the mind was also pursued by the Soviets for decades.

And, finally, we have this: "Visiting Soviet psi labs in 1967, Doctor Ryzl says he was told by a Soviet, 'When suitable means of propaganda are cleverly used, it is possible to mold any man's conscience so that in the end he may misuse his abilities while remaining convinced that he is serving an honest purpose.' Ryzl continues, 'The USSR has the means to keep the results of such research secret from the rest of the world and, as practical applications of these results become possible, there is no doubt that the Soviet Union will do so.' What will ESP be used for? 'To make money, *and as a weapon* [italics mine],' Ryzl states flatly." If such possibilities existed in 1972, that begs the question: Is ESP still being used to hypnotize people unknowingly? It's a disturbing question. As might very well be the answer.

Bear in mind that the document above was written way back in 1976. Imagine the extent of top-secret mind manipulation today.

Drugs and Altered Minds

Even more than mere mind manipulation, we see the slow, disturbing program designed to turn us further from being humans to robots. On April 13, 1953, CIA chief Allen Dulles ordered the creation of a program of mind control known as MKUltra to be conducted by Dr. Sidney Gottlieb. It was a program that grew to a huge size and encompassed research in such areas as hypnosis, the use of LSD for mind manipulation, hallucinogenics, chemical stimulation to control the human mind, and much more.

In the wake of the creation of MKUltra, rumors and half-truths about new mind-control techniques being used by Soviet, Chinese, and North Korean interrogators on U.S. prisoners of war panicked the CIA into a search for its own sure-fire method of questioning captives. In April 1961, Dr. Gottlieb decided his experiments with electrodes implanted in animals' brains were successful and that it was time to experiment with human brains. Information has leaked out concerning experiments with three Viet-Cong (VC) prisoners in July 1968.

A team of behaviorists flew into Saigon and traveled to the hospital at Bien Hoa, where the prisoners were being confined. The agents from Subproject 94 set up their equipment in an enclosed compound, and the team's neurosurgeon and neurologist inserted miniscule electrodes into the brains of the three VC prisoners. After a brief recovery period, the prisoners were armed with knives, and direct electrical stimulation was applied to their brains. The goal of the experiment was to determine if individuals with such electrodes implanted in their brains could be incited to attack and to kill one another. The agency was seeking a perfect sleeper assassin, a true Manchurian Candidate, who could be electronically directed to kill a subject. After a week of enduring electrical shocks to their brains, the prisoners still refused to attack one another. They were summarily executed, and their bodies burned.

In the 1959 novel The Manchurian Candidate *by Richard Condon, which was adapted into a 1962 film (shown here) and again adapted in 2004, a prominent member of a rich American family is brainwashed by the Chinese to become a political assassin.*

Such was the controversial nature of the work of MKUltra, the above example being just one of many, in 1973. Sidney Gottlieb—acting on the orders of CIA director Richard Helms—took quick and careful steps to ensure that all of the MKUltra files would be destroyed. Indeed, by 1973 the programs were years beyond the testing stage, so it was not seen as overly important to preserve the old research files of the 1950s and 1960s. The result was that an untold number of documents were shredded and destroyed under Gottlieb. Fortunately, a large amount of documentation survived destruction, largely because as there was such a massive amount of material that it was hard for even the CIA to determine where it was all stored. Some was in the hands of others agencies, and other stashes were held by companies to whom the CIA had contracted certain aspects of the MKUltra program. The result was that the CIA's mind-control files were all over the place, thus making it very difficult for CIA director Helms's plans to destroy all the files to go ahead successfully.

Today, MKUltra no longer exists. However, there is another form of mind control that is staring all of us in the face—although, admittedly, most people don't realize that it even falls into the mind-control domain. It's the issue of—*and the misuse of*—mood-altering drugs. Such drugs are now in widespread use throughout the United States, where they are used and abused to a greater level than in any other country on the planet. In other words, millions of Americans are under the constant sway of mind-manipulating drugs, specifically to curb the effects of issues such as anxiety, panic attacks, depression, obsessive compulsive disorder, and manic behavior. In the twenty-first century, government agencies do not need to use covert mind-control programs to try to control us. The population is doing it themselves—purposefully.

In the twenty-first century, government agencies do not need to use covert mind-control programs to try to control us. The population is doing it themselves—purposefully.

Global Research makes a very good point: "Here in the early stages of the twenty-first century, a ruling elite has manipulated our planet of seven billion people into a global economic system of feudalism.... By examining one aspect of this grand theft planet through the story of Big Pharma, one can accurately recognize and assess Big Pharma's success in its momentum-gathering power grab. Its story serves as a microcosm perfectly illustrating and paralleling the macrocosm that is today's oligarch engineered, highly successful New World Order nightmare coming true right before our eyes that we're all now up against."

This is not, however, just about money. It's the nature of the pharmaceuticals that we also need to focus on. In 2013, the *New York Times* noted something shocking: "Over the past two decades, the use of antidepressants has skyrocketed. One in 10 Americans now takes an antidepressant medication; among women in their 40s and 50s, the figure is one in four. Experts have offered numerous reasons. Depression is common, and economic struggles have added to our stress and anxiety. Television ads promote antidepressants, and insurance plans usually cover them, even while limiting talk therapy. But a recent study suggests another explanation: that the condition is being overdiagnosed on a remarkable scale."

One year later, *Scientific American* told its readers: "Antidepressant use among Americans is skyrocketing. Adults in the U.S. consumed four times more antidepressants in the late 2000s than they did in the early 1990s. As the third most frequently taken medication in the U.S., researchers estimate that 8 to 10 percent of the population is taking an antidepressant."

Americans are hooked on antidepressants and opioids. It might be, though, that they don't need them nearly as much as they are being prescribed.

Healthline, too, revealed disturbing statistics: "Opioids are a class of drugs known for their ability to produce a euphoric high, as well as a debilitating addiction. They are strictly regulated for these reasons and should only be prescribed to treat chronic pain resulting from a disease, surgery, or injury, according to the Center for Addiction and Mental Health. Opioids—codeine, oxycodone, and hydrocodone—are increasingly popular recreational drugs that have been both celebrated and condemned in popular culture.

Forbes noted that decades after Xanax was first placed on the market in 1981, "America is still a Xanax nation. It remains the most popular psychiatric drug, topping more recently introduced medicines like the sleeping pill Ambien (No. 2) and the antidepressant Lexapro (No. 3). Doctors write nearly 50 million prescriptions for Xanax or alprazolam (the cheap, generic equivalent) every year—that's more than one Xanax prescription every second.

While mood-altering drugs most assuredly help to combat anxiety-driven conditions, they are often so powerful that they create a near-zombified state in the user. For those who seek to control the population, it's a perfect win-win situation: those in the pharmaceutical industry who pull in billions and billions of dollars are happy to see more and more of the population turning into something close to zombies: millions of people fogged, sedated, and rendered into states that in many examples prevent them from functioning as normal human beings.

There is absolutely no doubt that the abuse of what are known as prescription opioids is growing at an alarming rate, and it has been doing so since the 1980s. Of course, we should not place everyone who takes opioids in the classification of abusers. Most people take their prescription meds in a responsible fashion, but that cannot be said for everyone. It's a startling fact that more than 20 percent of the U.S. population has, at one time or another, taken mood-altering drugs—many on a regular basis, and more than a few to a reckless and dangerous degree. In 2014, the National Survey on Drug Use and Health uncovered startling data: that within that same year, more than two million Americans had taken mood-altering drugs when they were not actually in need of medical help. Those two million–plus took such medications because they enjoyed the "high," the somewhat spaced-out feeling, that such meds provide as they calm down the user. The survey revealed that one of the biggest problems facing America was how such drugs—Prozac and Xanax, to list just two—are being used in "non-medical" ways. Abused, in other words. To put it into perspective: close to 6,000 people per day in 2014 chose to take prescription meds just for the hell of it. The primary groups were women, teenagers, and people in their twenties.

The National Institute on Drug Abuse (NIDA) says: "The reasons for the high prevalence of prescription drug misuse vary by age, gender, and other factors, but likely include ease of access. The number of prescriptions for some of these medications has increased dramatically since the early 1990s. Moreover, misinformation about the addictive properties of prescription opioids and the perception that prescription drugs are less harmful than illicit drugs are other possible contributors to the problem."

The NIDA also notes that when it comes to the nonmedical use of mind-altering drugs, it's the 18-to-25-year-old group that is most at risk from becoming hooked, overdosing, and even dying.

The NIDA also notes that when it comes to the nonmedical use of mind-altering drugs, it's the 18-to-25-year-old group that is most at risk from becoming hooked, overdosing, and even dying. For twelfth graders, only booze, cigarettes, and marijuana are ahead of the likes of Xanax and Prozac. There is some light at the end of the tunnel. Although teenagers are still using such drugs widely, figures have actually dropped appreciably in the last couple of years. Why that should be the case is not entirely clear. For the non-medical use of drugs among people 18 to 25 years old, the most popular drugs of choice are Adderall—which around one in 13 high-school seniors have taken—and Vicodin. The former is designed to lessen the effects of attention deficit hyperactivity disorder (ADHD), while Vicodin is a powerful pain reliever. Both, however, can give the user what amounts to a high—hence why they are abused when they are actually not needed by the user. Studies have shown that most young people get their drugs from friends, not via prescriptions or the internet. Sneaking a few pills from their parents' supplies is not, of course, unknown, but NIDA's research has shown that the sharing of supplies among friends is the primary way in which pills are acquired and used.

Interestingly, those in their teens and twenties who misuse mind-altering drugs are also likely to abuse alcohol—liquor and beer mostly, with wine at the bottom of the list—and drugs such as pot and cocaine. Now let's take a look at the situation in other age groups.

It may come as a surprise to many to learn that among those 55 to 80 years old, abuse of drugs is not at all unknown. In fact, it's a regular issue that NIDA has noted. Close to 85 percent of this group take one or two prescribed drugs per day, with more than a few exceeding what would be termed a safe dose. For the elderly, whose tolerance to powerful drugs is much less than that of younger, healthier people, overusing opiates can be a very dangerous game.

There are other noticeable revelations too: today, the number of young men who die from mood-altering prescription drugs is significantly higher than the number of young women who do so. It should be noted, though, that for the latter group, the figures are growing.

What all of this shows us is that the use of drugs that can have a major effect on the human brain is a major issue in the United States. We are becoming a nation in which millions of people are hooked on drugs that will not just give them a high but also a low: a relaxing of the mind, a lessening of anxiety, and a sense of calm. Some might say there is nothing wrong with that. On the other hand, one could make the disturbing observation that a nation that is overmedicated is also a nation that can easily be controlled, manipulated, and led toward a *Brave New World* type of future in which no one cares about much at all, beyond achieving the next high of their choice. The days of government-planned mind control are largely gone. The grim fact is that we are knowingly doing it to ourselves—and in the millions, too.

There's also the matter of our smartphones: the fact that just about everyone has one attached to themselves 24/7 practically already makes us cyborgs, albeit in a strange way. And the sinister side of this is that our technology secretly allows government agencies to snoop into just about every conversation we have. The number of people who own iPhones and other smartphones today is huge. Citing figures from Consumer Intelligence Research Partners (CIRP), *CNET* revealed in 2015: "An estimated total of 94 million iPhones were in use in the US at the end of March, including 38 million iPhone 6 and iPhone 6 Plus units, CIRP said Friday. That 38 million figure is made up of 25 million iPhone 6 units and 13 million iPhone 6 models, according to the company's estimates. CIRP also discovered that there are still 8 million iPhone 4S devices in use in the US, despite the handset launching in 2011.

"To put that in some context: Research firm eMarketer estimates 184.2 million people in the US will use a smartphone this year, representing 71.6 percent of mobile phone users and 57.3 percent of the population. By 2018, the firm predicts penetration will reach 82 percent of mobile phone users and 66.9 percent of the US population, with a total of 220 million smartphone users." (As of February 2021, 85 percent of Americans own smartphones.)

Smartphone ownership has reached deep penetration into American society, with 85 percent of Americans owning one as of 2021.

At the time this book was published, approximately six billion people had a smartphone. That means almost everyone had their every move and word open to governments. This is made all the more alarming by the fact that the US National Security Agency now has the ability to penetrate just about every iPhone on the planet—to the extent that your emails, social-media activity, and phone calls can all be carefully monitored 24/7. It's a dream come true for the New World Order. For us, it's a nightmare come true.

Andrew Griffin, writing at the *Independent* in 2015, said: "The iPhone has secret spyware that lets governments watch users without their knowledge, according to Edward Snowden. The NSA whistleblower doesn't use a phone because of the secret software, which Snowden's lawyer says can be remotely activated to watch the user. 'Edward never uses an iPhone, he's got a simple phone,' Anatoly Kucherena told Russian news agency RIA Novosti. 'The iPhone has special software that can activate itself without the owner having to press a button and gather information about him, that's why on security grounds he refused to have this phone.'"

If this all sounds like over-the-top paranoia, the bad news is that it's not. Griffin adds: "Apple has been active in making the iPhone harder for security services to spy on, and the company said that iOS 8 made it impossible for law enforcement to extract users' personal data, even if they have a warrant. The company has also been active in campaigning for privacy reform after the Snowden revelations, joining with Facebook and Google to call for changes to the law.

"But recently published files from the NSA showed that British agency GCHQ used the phones' UDIDs—the unique identifier that each iPhone has—to track users. While there doesn't seem to be any mention of such spying software in any of the revelations so far, a range of documents are thought to be still unpublished."

But how could such a thing happen? How are agencies able to achieve this blatant violation of our privacy? *Wired*'s Andy Greenberg explains the process—which is chillingly simple when one knows how to do it: "Security researchers posit that if an attacker has a chance to install malware before you shut down your phone, that software could make the phone *look* like it's shutting down—complete with a fake 'slide to power off' screen. Instead of powering down, it enters a low-power mode that leaves its baseband chip—which controls communication with the carrier—on.

"This 'playing dead' state would allow the phone to receive commands, including one to activate its microphone, says Eric McDonald, a hardware engineer in Los Angeles. McDonald is also a member of the Evad3rs, a team of iPhone hackers who created jailbreaks for the two

previous iPhone operating systems. If the NSA used an exploit like those McDonald's worked on to infect phone with malware that fakes a shutdown, 'the screen would look black and nothing would happen if you pressed buttons,' he says. 'But it's conceivable that the baseband is still on, or turns on periodically. And it would be very difficult to know whether the phone has been compromised.'"

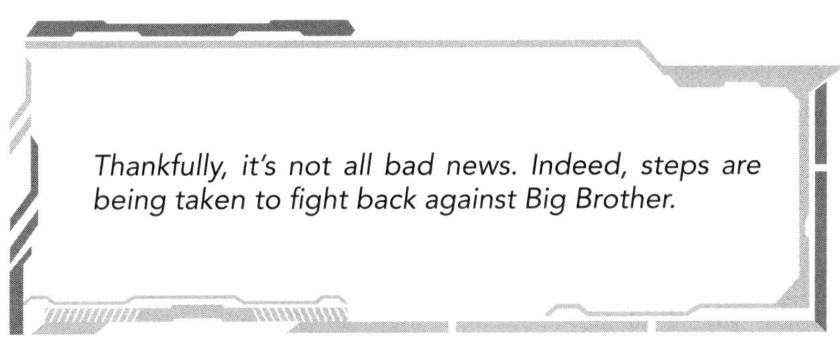

Thankfully, it's not all bad news. Indeed, steps are being taken to fight back against Big Brother.

Thankfully, it's not all bad news. Indeed, steps are being taken to fight back against Big Brother. CNBC says of this specific issue: "Stronger encryption in Apple's iPhones and on websites like Facebook has 'petrified' the U.S. government because it has made it harder to spy on communications, Glenn Greenwald, the writer who first reported on Edward Snowden's stolen files, told CNBC.

"Former National Security Agency (NSA) contractor Edward Snowden caused major shockwaves around the world in 2013 when he unveiled the surveillance body's wide ranging spying practices, which included regularly attempting to snoop of data held by major technology companies. Glenn Greenwald, the man who helped Snowden publish the documents, said that Silicon Valley companies have bolstered the encryption on their products, thereby making it harder for governments to eavesdrop."

Grown in Factories: A Grim Future

Of the many ways in which the human race might one day be "altered," surely the most nightmarish of all would for us to be genetically changed—"modified," we might say—and turned into something that will not be entirely human. At least, no longer the way that we understand the term "human" today. Something almost robotic, even. Is it possible that, one day, the tried and tested fashion for having a baby—a fashion that has worked very well for millions of years—will be no more? Incredibly, could it be the case that our children will be "designed"? Might they be endlessly reeled off on conveyor belts in giant factories? If the answers to those terrifying questions are a collective, decisive "yes," then might future humans be modified in ways that will see us dumbed down? The human equivalent of subservient cattle? Clever enough to work for our governments, but not astute enough to know that we are being treated like slaves?

The bad (and dangerous) news is that research is being undertaken right now that may well push us down the path of creating a human workforce that is good for nothing but, well, for work. And that workforce will be far more robotic than what we are today. If you think that such a thing could never, ever happen, you would be way off. It's already begun. And, it's likely not to stop.

On January 8, 2017, the *Guardian* newspaper revealed what it had learned on the latest news in the field of designing babies to order: "Novelist Kazuo Ishiguro, whose 2005 novel, *Never Let Me Go*, described children produced and reared as organ donors, last month warned that thanks to advances in gene editing, 'we're coming close to the point where we can, objectively in some sense, create people who are superior to others.'"

The *Guardian* asserted that even if such a situation does occur, then the likelihood is that it will not lead to the "engineering" of the populace but will be a means to have people sign up to the programs, thereby increasing profits of the respective companies involved rather than radically changing

Born in Japan and now a British citizen, Sir Kazuo Ishiguro is a Nobel Prize-winning author and musician. His novel Never Let Me Go *imagines a world in which children are bred merely to be organ donors. He has warned humanity about the dangers of gene editing.*

society. The *Guardian* urges restraint on all fronts when it comes to this controversial area of research. Nevertheless, not everyone is convinced that we will find ourselves heading down a road to a definitive nightmare. For example, a bioethicist named Henry Greely is of the opinion that the idea of creating either a "superman" or what he terms a "spilt in the species" is unlikely. More accurately, he doesn't see such a situation coming to fruition soon, mainly because "we don't know enough." That is, we don't know enough yet, but perhaps one day we will.

Certainly, the term "modified babies"—as the *Guardian* puts it—should be enough to ensure that we never go down a path that sees us becoming something less than we are—something that can be controlled and regulated. Of course, everyone should encourage medical advances, but turning significant portions of the human population into subservient, unquestioning slaves is hardly what the vast majority of people would term a positive prospect.

The *Guardian* admits that "every new advance puts a fresh spark of life into Huxley's monstrous vision." Certainly, the warnings and concerns of Kazuo Ishiguro are all too real. There is no doubt that many of Ishiguro's deep concerns were prompted by a procedure known as CRISPR-Cas9. CRISPR stands for clustered regularly interspaced short palindromic repeats, and Cas9 is the name of a specific protein that plays a role in defending the body against diseases. It's a process that falls under the admittedly sinister-sounding heading of "gene editing."

The process came into being in 2012, and it was created to find ways to eradicate so-called "mutant genes" that can wreak havoc in the human body and lead to severe illness and death. In that sense, the process is a positive one. On the other hand, however, as is so often the case in such situations, the possibility of crossing the line and abusing the science and technology involved cannot be ruled out. What we are faced with, then, is a definitive double-edged sword—one that offers a future free of disease but, if misused, may take us down a path in which only nightmares can be found.

Human gene editing was once the purview of science fiction. It is now a reality that comes with many serious moral and evolutionary implications.

That we may already be heading down that dark path is not just a theory. For example, since 2012, CRISPR-Cas9 has been used in China as a means to "modify" human embryos. So far, the results of such tinkering and tampering with the natural order have produced nothing but mixed results. The United Kingdom is getting on board with CRISPR-Cas9, too: the U.K.'s Francis Crick Institute has been issued a license to allow it to use CRISPR-Cas9 on two-to-three-day-old human embryos, chiefly to try to understand some of the reasons why miscarriages occur.

The biggest problem with all this is not just the fact that the technology exists and is ripe for manipulation of a sinister kind. No; the major issue revolves around the fact that throughout the world, hard and definitive legislation does not exist to prevent scientists from going down certain roads they have no business going down. Since it's an area that is filled with unknowns, it's equally unknown how exactly we should proceed. Of course, that is all understandable, but it's also an issue that needs far more research—specifically to ensure that 100 years from now, we are not all under the control of crazed scientists whose mandate is to create a superior class of people and an inferior one. The rise of a real master race? Even the *Guardian*, which sees the positive aspects of the technology, admits that unchecked developments might push us down "a path towards non-therapeutic genetic enhancement."

In January 2015, the BBC immersed itself in the controversial domain of "designer babies." It warned that experts in the field of advanced genetics were coming to a stark realization that society needs to "be prepared" for what the future may bring. One of those was Dr. Tony Perry, one of the foremost people in the field of cloning. When he stated that we're no longer talking about "H. G. Wells territory," he meant that science fiction was rapidly becoming science fact. It's hardly surprising that cries to curb certain programs, until the implications of such research could be fully ascertained, were heard here, there, and everywhere.

In a genetically engineered future, would humans all become blonde, blue-eyed, healthy but obedient automatons for a "pure society"?

Dr. Perry most definitely knows what he is talking about. He was, for example, one of the very first people to successfully clone animals, such as pigs and mice. In an article for the publication *Scientific Reports*, said the BBC, Dr. Perry "details precisely editing the genome of mice at the point DNA from the sperm and egg come together."

Dr. Perry explained for the BBC: "We used a pair of molecular scissors and a molecular sat-nav that tells the scissors where to cut. It is approaching 100% efficiency already, it's a case of 'you shoot you score.'"

Two years later, in 2017, the *Blaze* turned its attention to these controversies, too. The publication noted that a group within the National Academy of Sciences and National Academy of Medicine was "advocating for 'germ-line modification' of human babies in certain narrow circumstances to prevent the birth of children with serious diseases." The *Blaze* got to the crux of the matter with lightning speed, noting that the biggest worries concerned the abuse of technology "for modifying the germ-line—or inherited DNA—of human beings because it could lead to 'designer babies' with pre-selected eye color, physical strength or even intelligence."

One of those who urged caution on the introduction of such science—a form of science that could easily, one day, be turned against us by controllers—was Dr. Marcy Darnovsky, whose concerns the *Blaze* noted. A "liberal policy advocate" based at the Center for Genetics and Society, Dr. Darnovsky said at the dawning of the twenty-first century that there was a very real possibility that such technology could "significantly exacerbate socio-economic inequality." The result may well see only the rich having designer babies, which would widen the chasm between "the upper and lower classes."

In February 2017, *Industry Leaders* pointed out: "Organizations like 23andMe and GenePeeks, Inc. are receiving backlash from the media and from organizations dealing with reproductive issues." Some of that backlash came from Dr. Darnovsky. She was very vocal on this particular issue and said: "It would be highly irresponsible for 23andMe or anyone else to offer a product or service based on this patent. It amounts to shopping for designer donors in an effort to produce designer babies. We believe the patent office made a serious mistake in allowing a patent that includes drop-down menus from which to choose a future child's traits."

Industry Leaders added: "Eventually, 23andMe succumbed to the pressure and wrote off its inheritance calculator service in fertility treatments."

From the White House to COVID-19

To demonstrate the extent to which we are becoming a robot world, let's focus on two issues of importance: the office of the White House and COVID-19. In 2011, the White House announced the following:

"One exciting element of the President's Advanced Manufacturing Partnership is the National Robotics Initiative. Robots are working for us every day, in countless ways. At home, at work, and on the battlefield, robots are increasingly lifting the burdens of tasks that are dull, dirty, or dangerous.

"But they could do even more, and that's what the National Robotics Initiative is all about. So today, four agencies (the National Science Foundation, the National Institutes of Health, NASA, and the United States Department of Agriculture) are issuing a joint solicitation that will provide up to $70 million in research funding for next-generation robotics.

"The focus of this initiative is on developing robots that work with or beside people to extend or augment human capabilities, taking advantage of the different strengths of humans and robots. In addition to investing in the core technology needed for next-generation robotics, the initiative will support applications such as robots that can:

- Increase the productivity of workers in the manufacturing sector;

- Assist astronauts in dangerous and expensive missions;
- Help scientists accelerate the discovery of new, life-saving drugs; and
- Improve food safety by rapidly sensing microbial contamination.

"The initiative will also accelerate progress in the field by requiring researchers to share the software and robotics operating systems they develop or contribute to and funding the purchase of robotics platforms.

"The Administration has decided to make robotics a priority because:

Robots aren't just used in places like factories and in outer space. For example, here is an agricultural robot that employs AI to control weeds, add fertilizers, and even harvest fruits and vegetables.

- Robotics can address a broad range of national needs such as advanced manufacturing, logistics, services, transportation, homeland security, defense, medicine, healthcare, space exploration, environmental monitoring, and agriculture;

- Robotics technology is reaching a 'tipping point' and is poised for explosive growth because of improvements in core technologies such as microprocessors, sensors, and algorithms;

- Robotics can play an important role in science, technology, engineering and mathematics (STEM) education because it encourages hands-on learning and the integration of science, engineering, and creative thinking; and

- Members of the research community such as the Computing Community Consortium and program managers in key sciences have developed a shared vision and an ambitious technical agenda for developing next-generation robotic systems that can safely work with humans and augment human capabilities.

"We want to thank the team of agency program managers that worked on development of this solicitation. We also want to encourage leaders in industry and academia to partner with the Administration as we work to promote U.S. leadership in next-generation robotics and its applications."

The National Science Foundation describes one of its own robotics initiatives as follows: "The National Robotics Initiative 3.0: Innovations in Integration of Robotics (NRI-3.0) program builds upon the preceding National Robotics Initiative (NRI) programs to support fundamental research in the United States that will advance the science of robot integration. The program supports research that promotes integration of robots to the benefit of humans including human safety and human independence.

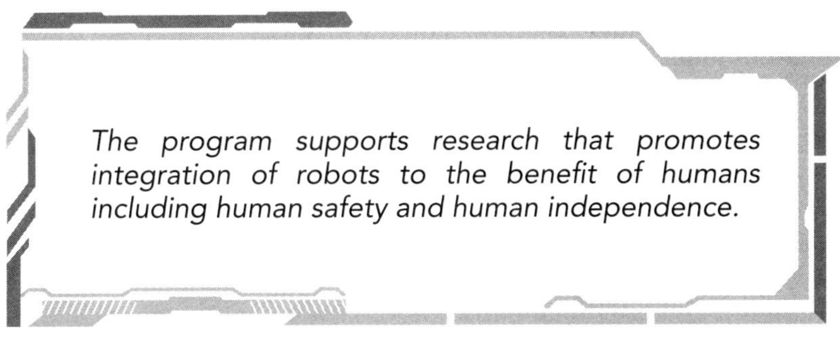
The program supports research that promotes integration of robots to the benefit of humans including human safety and human independence.

"Collaboration between academic, industry, non-profit, and other organizations is encouraged to establish better linkages between fundamental science and engineering and technology development, deployment, and use.

"The NRI-3.0 program is supported by multiple agencies of the federal government including the National Science Foundation (NSF), the U.S. Department of Agriculture (USDA), the National Aeronautics and Space Administration (NASA), the Department of Transportation (DOT), the National Institutes of Health (NIH), and the National Institute for Occupational Safety and Health (NIOSH). Questions concerning a particular project's focus, direction, and relevance to a participating funding organization should be addressed to that agency's point of contact, listed in section VIII of this solicitation."

Now let's investigate the relationship between robotics and COVID-19. The U.S. Department of Defense has made it clear that robotics can play a big role in the fight against the disease. The DoD provided the following in October 2020:

"A robot initially designed for shipboard firefighting and maintenance tasks has now been enlisted in the fight against COVID-19. The decontamination robot was funded by the Office of Naval Research and designed by several local universities. It was recently tested in Richmond, Virginia. The robot has four wheels and a mechanical arm that uses short-wave ultraviolet light to decontaminate surfaces. The current version requires humans to oversee and 'drive' the robot, but the hope is that the robot will become fully autonomous.

"'The value of robots to deploy UVC lamps for decontamination is that you can reduce exposure of humans to the UVC light, and the robot can reposition the lamps over surfaces you wish to decontaminate using its arms,' Dr. Thomas McKenna, a program officer in ONR's [Office of Naval Research's] Warfighter Performance Department said. 'When the robot was designed, there was no COVID[-19], but the combination of mobility and manipulation are a good match to this task.'

"ONR has sponsored fundamental research in human–robot interaction and humanoid robots for more than 20 years. This research is occurring at several universities, small companies and at the U.S. Naval Research Laboratory. ONR has provided research, as well as programmatic direction and resources. The NRL is advancing state-of-the-art robotics and human interaction with robots.

"The ONR robotics program is focusing on removing humans from dangerous situations. Additionally, the program wants robots to perform routine tasks, in order to free up sailors and Marines to concentrate on their primary responsibilities and training.

"The ONR program in humanoid robots has focused mainly on shipboard damage control, including firefighting and shipboard maintenance.

"'The robots that were developed at Virginia Tech and now at the University of Virginia were designed for shipboard firefighting and maintenance tasks,' McKenna said.

"'When COVID-19 emerged, I asked the NRL artificial intelligence lab to track how robots were being used around the world to help fight the disease, including their use in hospitals and for decontamination.'

"McKenna added that Dr. Tomonari Furukawa, a professor at the University of Virginia, recognized how he could improve robot decontamination by the use of the manipulators on his mobile robots, to adaptively position UVC lamps to illuminate surfaces like table-tops and chairs.

"The robots in this recent demonstration were teleoperated by students.

"'The robot can already disinfect by teleoperation. In August, we tested and successfully demonstrated disinfection of a room of COVID-19 at a testing center while teleoperating from a different building,' Furukawa said. 'We are currently developing the ability to build a 3D map showing disinfected surfaces, and possibly infected surfaces, with which we can next introduce autonomous disinfection. Full completion will need much more work, but we are planning to complete the first installation of the mapping capability by the end of this year.'

"Mobile robotics is still in an early stage, experts say, and more research is needed to make the robots more agile and more autonomous.

"'I was pleased at the control and mobility that was shown in the demo,' McKenna said. 'Of course, one would have to validate that the viral load was actually reduced on the exposed surfaces, and that was not a part of this demo. The mobility capabilities demonstrated are very promising as we move further in the development stage.'"

Closing with a Collection of Oddities

Now we will investigate a series of issues that are all relevant to the world of the robot and of the manipulation of the human race. Quite possibly, the closest real-world equivalent to the fictional Dr. Frankenstein of Mary Shelley's classic novel of 1818, *Frankenstein*, was a Russian scientist who had a disturbing fascination with the idea of hybridizing different kinds of animals. It was a fascination that bordered on dangerous, crazed obsession. The man's name was Ilya Ivanov, and he entered this world in August 1870 in the Russian town of Shchigry. He was not destined to remain in his small-town environment, however. Ivanov obtained his professorship in 1907 and took employment at an animal sanctuary located in the Ukraine, specifically the province of Kherson Oblast. It was here that Ivanov's dark and disturbing research began.

Ivanov's initial work was focused on horses. He developed the idea of interbreeding racehorses to the point where, eventually, he would end up with what we might call the ultimate "superhorse," one that could outrun and outperform just about any other horse on the planet. His program, as one might imagine, did produce fine racehorses, but they were nothing out of the ordinary. Something that was very much out of the ordinary was Ivanov's next target of interest: fusing apes and humans into one.

Just a few years after the "superhorse" saga caught the attention of the Russian government's Academy of Science, Ivanov was given a sizeable grant by that academy to research the feasibility of creating something akin to an ape-man. It was a monstrous, abhorrent idea—but that didn't stop it from proceeding. Indeed, significant funding was provided to Ivanov, as was a wealth of medical equipment and even newly constructed laboratories where the terrible experiments could take place. The labs were not situated in the heart of the Soviet Union, however. No; the huge funding allowed for the construction of several facilities in West Africa, specifically in Kindia, Guinea. Thus, thanks to the help of the Pasteur Institute, the work began in early 1926.

Within months, Ivanov was delving into extremely controversial areas, as Stephanie Pain notes: "Ivanov passed the summer in Paris, where he spent some of his time at the Pasteur Institute working on ways to capture and subdue chimps, and some with the celebrated surgeon Serge Voronoff, inventor of an increasingly fashionable 'rejuvenation therapy.' In a now notorious operation, Voronoff grafted slices of ape testes into those of rich and ageing men hoping to regain their former vigor. That summer, he and Ivanov made headlines by transplanting a woman's ovary into a chimp called Nora and then inseminating her with human sperm."

Ivanov's team utilized both chimpanzees and gorillas in their experimentation—all of which was focused upon trying to successfully impregnate ape and chimpanzee females with human male sperm. In one regard it made great sense: chimpanzees, for example, have a DNA sequence that is around 95 percent identical to that of the human race. What seemed promising on paper and in theory, however, proved to be far less promising in reality. That is to say, none of the attempts at impregnation worked. Ivanov was not dissuaded, however. He decided to take a different approach—a very dangerous and even unethical approach.

A Soviet biologist and expert on artificial insemination, Ilya Ivanov (1870–1932) tried to crossbreed apes with human beings.

If the process of impregnating apes with human sperm wasn't going to work, thought Ivanov, why not impregnate human women with ape sperm? This is precisely what Ivanov did—on local women. It seems, from the surviving records, that little if any thought was given to (a) the moral aspects of all this (or, to be correct, the profound lack of morals); and (b) the trauma that the women might experience in the event that they gave birth to freakish half-human, half-ape abominations. Despite this approach, and regardless of whether the animals used were gorillas or chimpanzees, failure was the only outcome.

Ivanov was not still deterred. He returned to his native Soviet Union, secured further funding, and established several other labs. Rather intriguingly, some of the locations of the installations were kept secret—perhaps to ensure that outraged locals didn't take to storming them, fearful of what was afoot in their towns and cities. Nevertheless, it is a fact that at least one facility was built—underground—in the town of Sukhumi, Georgia. We know this because, just a few years ago, a number of ape skeletons were unearthed there when workmen digging underground stumbled upon one of the nightmarish labs.

The Soviet government was hardly pleased with the outcome, and under the ruthless command of Joseph Stalin, its agents arrested Ivanov.

It is perhaps not a surprise to learn that the experiments undertaken in Sukhumi failed as abysmally as those attempted in Guinea. The Soviet government was hardly pleased with the outcome, and under the ruthless command of Joseph Stalin, its agents arrested Ivanov. It was December 31, 1932, when Ivanov found himself in distinctly hot water. It soon became scalding: he received a five-year term in prison and died just two years into his sentence, which had been changed from prison to exile shortly after the sentence.

Pawel Wargan, who has extensively studied the work of Ivanov, notes: "Interspecific hybridization was seen to hold great potential. Animals that combined the strongest qualities of two species could become popular house pets. The Soviet media was keen to suggest that a new species, uniting human strength with the subservience and agility of an ape, could form a more obedient workforce, a stronger army. The Soviet Union was caught in a genetic manipulation mania, much to the amusement of one novelist—[Mikhail] Bulgakov wrote of a canine that became a Soviet bureaucrat after being subject to a transplant of human testicles. The buildings on this hill above Sukhumi were to be the Soviet answer to Darwin's insights, where chimeras were born and biology became another tool in the propagandist's arsenal."

The final word goes to Jerry Bergman, Ph.D., who says: "In the end, the research failed and has not been attempted again, at least publicly. Today we know it will not be successful for many reasons, and Professor Ivanov's attempts are, for this reason, a major embarrassment to science. One problem is humans have 46 chromosomes—apes 48—and for this reason the chromosomes will not pair up properly even if a zygote is formed. Another problem is a conservatively estimated *40 million* base pair differences exist between humans and our putative closest evolutionary relatives, the chimps. These experiments are the result of evolutionary thinking and they failed because their basic premise is false."

Achieving immortality is something that just might be achieved via becoming a cyborg, but there are other ways, too. None of them have worked yet, however.

Significant portions of this overall story revolve around immortality—or, rather, a quest for immortality. Achieving immortality is something that just might be achieved via becoming a cyborg, but there are other ways, too. None of them have worked yet, however. One such example is that which revolves around the notorious saga of Heaven's Gate, as we shall soon see. Life is something that we experience until we no longer do so. Depending on your personal beliefs, death is either a state of never-ending "lights out" or the start of a new and endless adventure. But what about reanimation and resurrection? Maybe there's a chance of that, if you're half-human and half-robotic. The notion is not for everyone, though, as the following story demonstrates.

One person who was convinced that dying and coming back—in some form, at least—would be a wholly positive experience was a controversial character named Marshall Herff Applewhite Jr.

A native of Texas, Applewhite gained infamy in 1997 when he convinced 38 of his followers in the so-called Heaven's Gate cult to take their own lives—chiefly because doing so would see them return in reanimated, immortal form. But before we get to death, reanimation, and immortality, let's see what it was that led to that terrible tragedy.

From a very young age, Applewhite's life was dominated by religious teachings: his father was a minister who lectured to, even thundered at, his meek followers. It was made clear to Applewhite that he was expected to do likewise. Exhibiting a high degree of youthful rebellion, however, Applewhite did not. Instead, he joined the U.S. Army. After leaving the military (which involved him spending a lot of time at the White Sands Proving Ground, New Mexico), Applewhite took his life in a very different direction: he became a music teacher.

He could not, however, shake off altogether that religious programming he had received as a child and took a job with St. Mark's Episcopal Church in Houston, Texas. Demonstrating that he was not quite so saintly after all, however, in 1974 he was arrested for credit card fraud. It was also in the 1970s that Applewhite met the love of his life, Bonnie Nettles. From then on, the devoted pair delved more and more into the world of cultish, crackpot activity: they established the Total Overcomers Anonymous (TOA) group, which assured its followers that benevolent aliens were out there, ready and willing to help all those who pledged allegiance to the TOA. And many did exactly that. It was a group that eventually mutated into the infamous Heaven's Gate cult. And this brings us up to March 1997.

Three weeks into the month, Applewhite started brainwashing his duped clan into believing that if they killed themselves, they

Three weeks into the month, Applewhite started brainwashing his duped clan into believing that if they killed themselves, they would all reanimate—not here, but in some angelic dimension far different to, and far away from, our own earthly realm.

would all reanimate—not here, but in some angelic dimension far different to, and far away from, our own earthly realm.

Since the comet Hale-Bopp was just around the corner, so to speak, Applewhite even weaved that into his story. There was, he said, a huge UFO flying right behind the comet, and when death came for the group, the UFO would transfer them aboard the craft and to new and undead lives elsewhere.

His loyal followers eagerly swallowed every word. To their eternal cost, they eagerly swallowed something else, too: highly potent amounts of phenobarbital and vodka. In no time at all, almost 40 people were dead, all thanks to the words of Applewhite. Aliens did not call upon the members of the Heaven's Gate group. No UFO was ever detected behind Hale-Bopp. And the dead did not rise from the floor of the Heaven's Gate abode, which was situated at Rancho Santa Fe, California, and where one and all took their lives. Their bodies stayed exactly where they were until the authorities took them to the morgue for autopsy.

There is a major lesson to be learned here: don't base your life around the claims of a man who tells you that knocking back large amounts of phenobarbital and vodka will ensure your reanimation and immortality. It won't. Maybe, one day, becoming a cyborg will provide the answer.

It may intrigue many to learn that there exists a medical condition called walking corpse syndrome (WCS). For the sufferers, it's an absolute nightmare and no joke at all. There is a reason why I mention it: it's because some of the sufferers believe that their limbs are not theirs. In other words, they are replacement limbs. Even stranger, some believe that those limbs aren't even made of flesh—that they only look that way. You can see how that can easily slide into the world of robots and cyborgs.

WCS is officially known as Cotard's syndrome. It's a condition steeped in mystery and intrigue. Disturbingly, Cotard's syndrome causes the victim to believe that he or she is dead or that their limbs are no longer living or even theirs. The condition takes is name from one Jules Cotard, a French neurologist who died in 1889 from diphtheria. He spent much of his career studying and cataloging cases of walking corpse syndrome. Not only do those affected by WCS believe they are dead, but they also fall into spirals of psychosis and fail to take care of their personal appearance.

One of the most disturbing cases of walking corpse syndrome surfaced in the United Kingdom in May 2013. A man, referred to by the medical community only as "Graham," found himself descending into a deeply depressed state, to the point where he ultimately came to believe he was literally a member of the walking dead club. As Graham's condition rapidly worsened and he actually spent his days and nights wandering around graveyards, his family was forced to seek medical treatment. For a while, Graham became convinced that his

Dr. Jules Cotard (1840–1889), a French neurologist, is most often remembered for "Cotard's syndrome," the rare belief that one is dead.

brain was either clinically dead or was "missing" from his skull. Fortunately, treatment finally brought Graham back to the world of the living.

New light was soon shed on the nature of Cotard's syndrome as a result of a connection to Zovirax, generally used in the treatment of herpes-based conditions, such as cold sores. Although Zovirax is known for having a small number of side effects, studies revealed that approximately 1 percent of people prescribed Zovirax developed psychiatric conditions, including Cotard's syndrome. Intriguingly, most of those who were taking Zovirax and experienced walking corpse syndrome were suffering from renal failure at the time.

Studies undertaken by Anders Helldén of Karolinska University Hospital in Stockholm, Sweden, and Thomas Linden of the Sahlgrenska Academy in Gothenburg, Sweden, have uncovered remarkable, albeit unsettling, data on this curious phenomenon. Their case studies included that of a woman who was prescribed Zovirax after having a bout of shingles. When the drug took hold of the woman, who also happened to have renal failure, she began to act in a crazed and concerned fashion, suspecting that she was dead. When given emergency dialysis to cope with the effects of kidney failure, her strange beliefs began to fade, to the point where she finally came to accept that she was not dead, after all. For hours, however, she remained convinced that "my left arm is definitely not mine."

The *Independent* said of this strange saga: "The woman ran into a hospital in an extremely anxious state, author of the research Anders Helldén from the Karolinska University Hospital in Stockholm said. After receiving dialysis, the woman explained that she had felt anxious because she had been overwhelmed by a strong feeling that she was dead. Within a few hours her symptoms began to ease, until she felt that she was 'pretty sure' she wasn't dead but remained adamant her left arm did not belong to her. After 24 hours, her symptoms had disappeared. Blood analysis later revealed that acyclovir, which can normally be broken down in the body before being flushed out by the kidneys, can leave low levels of breakdown product CMMG in the body. Blood tests of those who had Cotard's symptoms showed much higher levels of CMMG. All but one of those tested also had renal failure."

35

Conclusions

So, with our story now over, what can we say about the world of robots? Well, we can say for sure that robots have played a collective, significant role in the history of the human. And that role goes back millennia to the work of Archimedes. In that sense, there's nothing new about robots; it's just a case of how sophisticated they were then versus now. Indeed, even in the early days, things were already startlingly advanced. After all, giant iron claws and laser beams would not be out of place in a 1950s-era sci-fi movie, never mind in the ancient history of the human race. The same goes for the Golem and the Homunculus: both were ancient, created humanoids. Then there is the way in which robots played a strange—almost bizarre—role in World War II and the Cold War. After all, who could have imagined that Jasper Maskelyne's "robot scarecrow" would have played such a decisive role in the war against Hitler's cronies? Very much the same goes for the Flatwoods Monster—a robot designed and dispatched to scare the hell out of people. And we should not forget the grisly *Acoustic Kitty*, a program of the Central Intelligence Agency that had one goal only in mind: to spy on the Russians by turning a normal cat into something tragically awful.

It was in this period—the post–World War II era—when the robot really came to the fore in the field of sci-fi, particularly in the world of the cinema: *The Day the Earth Stood Still* and *Forbidden Planet* are two perfect examples. Both entertained people and made them think, too. And in the 1960s, things really began to take off in the domain of the robot. As we've seen, a great deal of the development of the robot was provoked yet again by the world of the military—sometimes for good and sometimes not so good, as we'll see now.

As has been demonstrated, there have been absolutely incredible leaps and bounds in the developments of artificial limbs, many of which were made for warriors who were injured on battlefields. In that sense, the robot has done a great job. We're talking about "bionic" eyes and even new legs and arms. Also, there are the plans of the U.S. military to have cyborgs fighting on our battlefields by 2050. The same goes for the police: they're coming close to being real-life Robocops. Handled the right way, of course, these developments are perfect technologies for the world of the future. Handled wrongly, however—such as having robots that become bullying, controlling monsters—these developments could see us plunged into a veritable nightmare.

And there's yet another angle to all of this. I should stress that just about all of the robot technology that has been highlighted in the pages of this book is not just incredible but at times mind-boggling. We should, all of us, embrace such technology. After all, going backward or standing still does nothing for any of us. The future is the way for us, as it should be. But as we progress, we need to be aware of the fact that there is a dark side to all of this. It relates to how we, the human race, might—bit by bit—become the robots and cyborgs that we created in the first place. I am, of course, talking about the designing of what we might term "created humans."

Clones, modified humans, designer babies, half-human robots, and half-human astronauts might sound absolutely amazing; indeed, they would be. There is, however, one big concern we should all take notice of: changing people and altering babies to order might not necessarily be a positive thing. In a very short period of time, we might see a radical change in what passes for human and what passes for robot. I don't exaggerate at all when I say that the lines might blur quickly—*very* quickly.

Then there is the matter of sex. As we have seen, the matter of sex with robots, rather than with other people, is growing. A lot of people are going for it, and a couple of decades from now, sex with robots may be the norm. By then, that incredible creation named Sophia may just be seen as a quaint antique—although I'm sure she wouldn't like me to say that!

There is a further situation, too: it's the one that revolves around not our created robots, but robots from other worlds. Maybe other galaxies. We've seen how the extraterrestrial robot has played significant roles in TV shows and in the cinema, but there are also those alien, biological robots that Philip Corso talked about: the robotic devices designed to wipe us out—or so whistleblower Dan Salter claimed. Even weirder and creepier, there are the Women in Black, the Men in Black, and the Black-Eyed Children—three groups that seem to have what we might call "robotic aspects" attached to them. And, at the extreme, there are those highly controversial claims of people—specifically politicians and famous people—being replaced in a fashion not unlike the 1975 film *The Stepford Wives*.

Another huge potential danger is that of the purported "robot assassins." President John F. Kennedy, Martin Luther King Jr., Robert Kennedy, and John Lennon all may have been killed by mind control–driven figures resembling nothing less than remotely controlled robots. The results of MKUltra-type technology? We should not dismiss such a possibility. One only has to look at all four deaths to see how there was far more to things than meets the eye.

What all of this tells us is that the robot is a "creature" that we need to deal with carefully. There was a time when we were the greater power. Now, that may not be the case. A few decades from now, the robots could well be running the show. That is, unless we do the right thing: namely, to ensure that there is a perfect amount of balance in the worlds of science and technology. As I said, we should not keep away from the future world of robotics, but we should make sure we don't end up as mind-controlled slaves or designer humans. The way things are going, that could just happen. Unless, that is, balance becomes the order of the day. I have to say, though, I'm not at all sure such balance will be achieved. I hope that, 20 years from now, I won't be saying "I told you so." I have a nagging concern, however, that that's *exactly* what I'll be saying.

Further Reading

Adams, Tom. *The Choppers—and the Choppers: Mystery Helicopters and Animal Mutilations.* Paris, TX: Project Stigma, 1991.

Adamski, George. *Behind the Flying Saucer Mystery*. New York: Paperback Library, 1967.

Andrews, Bill. "If Wormholes Exist, Could We Really Travel through Them?" *Discover,* July 30, 2019. https://www.discovermagazine.com/the-sciences/if-wormholes-exist-could-we-really-travel-through-them.

Angelucci, Orfeo. *The Secret of the Saucers.* Stevens Point, WI: Amherst Press. 1955.

Associated Press. "Family Implanted with Computer Chips." *USA Today,* May 10, 2002. http://usatoday30.usatoday.com/tech/news/2002/05/10/implantable-chip.htm.

Ball, Philip. "Designer Babies: An Ethical Horror Waiting to Happen?" *The Guardian,* January 8, 2017. https://www.theguardian.com/science/2017/jan/08/designer-babies-ethical-horror-waiting-to- happen.

"Barcode Everyone at Birth." BBC.com, November 18, 2014. http://www.bbc.com/future/story/20120522-barcode-everyone-at-birth.

Barker, Gray. *M.I.B.: The Secret Terror Among Us.* Clarksburg, WV: New Age Press, 1983.

———. *They Knew Too Much About Flying Saucers.* New York: University Books, 1956.

Bartlett, Jamie. "Meet the Transhumanist Party: 'Want to Live Forever? Vote for Me.'" *The Telegraph,* December 23, 2014. http://www.telegraph.co.uk/technology/11310031/Meet-the-Transhumanist-Party-Want-to-live-forever-Vote-for-me.html.

Beckley, Timothy Green. *The UFO Silencers (Special Edition).* New Brunswick, NJ: Inner Light Publications, 1990.

Beckley, Timothy Green, and John Stuart. *Curse of the Men in Black.* New Brunswick, NJ: Global Communications, 2010.

Bellis, Mary. "Who Pioneered Robotics?" ThoughtCo, July 3, 2019. https://www.thoughtco.com/timeline-of-robots-1992363.

Bender, Albert. *Flying Saucers and the Three Men.* New York: Paperback Library, 1968.

Benitz, Jennifer. "Changing the Way Soldiers Fight and Survive: AI on the Battlefield." Association of the United States Army, January 24, 2020. https://www.ausa.org/articles/changing-way-soldiers-fight-and-survive-ai-battlefield.

Bishop, Katie. "Sex Robots, Teledildonics, and the Rise of Technosexuals During Lockdown." *Observer,* October 21, 2020. https://observer.com/2020/10/sex-robots-teledildonics-growing-popularity-covid19/.

Bostrom, Nick. "Human Genetic Enhancements: A Transhumanist Perspective." *Journal of Value Inquiry*, Vol. 37, No. 4, pp. 493–506.

Bowart, Walter. *Operation Mind Control.* New York: Dell Publishing, 1978.

Breslin, Susannah. "Meet the Terrifying New Robot Cop That's Patrolling Dubai." *Forbes,* June 3, 2017. https://www.forbes.com/sites/susannahbreslin/2017/06/03/robot-cop-dubai/?sh=2cf754c76872.

Brocklehurst, Steven. "The UFO Sighting Investigated by the Police." BBC.com, November 9, 2019. https://www.bbc.com/news/uk-scotland-50262655.

Brodeur, Nicole. "Sex Robots: An Answer for Aging, Lonely Americans in the Age of AI?" *Seattle Times,* November 17, 2020. https://www.seattletimes.com/seattle-news/sex-robots-an-answer-for-old-and-lonely-in-the- age-of-ai/.

Brown, Larisa. "Military Unveils Insect-Sized Spy Drone with Dragonfly-Like Wings." *Daily Mail,* August 11, 2016. http://www.dailymail.co.uk/sciencetech/article-3734945/Now-bugged-Military-unveils-insect-sized-spy-drone-dragonfly-like-wings.html.

Chrysopoulos, Philip. "Talos: The Giant Mythological Robot Protecting Crete." Greek Reporter, May 6, 2021.https://greekreporter.com/2021/05/06/talos-ancient-greek-robot/.

Clark, Josh. "What Was Archimedes' Death Ray?" Howstuffworks. Accessed March 20, 2021. https://history.howstuffworks.com/historical-figures/archimedes-death-ray.htm.

Cofield, Calla. "Time Travel and Wormholes: Physicist Kip Thorne's Wildest Theories." Space.com, December 19, 2014. https://www.space.com/28000-physicist-kip-thorne-wildest-theories.html.

Cope, Alec. "The Commercial That Shows What TV Does to Your Brain." Collective Evolution, November 19, 2014. http://www.collective-evolution.com/2014/11/19/this-is-what-tv-can-do-to-your-brain/.

Denison, Caleb. "Samsung Smart TVs Don't Spy on Owners." Digital Trends, February 9, 2015. http://www.digitaltrends.com/home-theater/samsung-tvs-arent-spying-eavesdropping- listening/.

Duca, Marc Di. "Legend of the Golem of Prague." Culture Trip, July 2, 2019. https://theculturetrip.com/europe/czech-republic/articles/the-legend-of-the-golem-of-prague/.

"First Human Clone Born, Cult Chemist Claims." FoxNews.com, December 27, 2002. http://foxnews.com/story/2002/12/27/first-human-clone-born-cult-chemist-claims.html.

Fryer-Biggs, Zachary. "Coming Soon to a Battlefield: Robots That Can Kill." *The Atlantic,* September 3, 2019. https://www.theatlantic.com/technology/archive/2019/09/killer-robots-and-new-era-machine-driven-warfare/597130/.

Gallagher, James. "'Designer Babies' Debate Should Start, Scientists Say." BBC.com, January 19, 2015. http://www.bbc.com/news/health-30742774.

Garrison, Jim. *On the Trail of the Assassins.* London, UK: Penguin Books, 1988.

Gent, Edd. "A New Bionic Eye Could Give Robots and the Blind 20/20 Vision." SingularityHub, May 22, 2020. https://singularityhub.com/2020/05/22/a-new-bionic-eye-could-give-robots-and-the-blind-20- 20-vision/.

Ghosh, Pallab. "Sex Robots May Cause Psychological Damage." BBC.com, February 15, 2020. https://www.bbc.com/news/science-environment-51330261.

Glod, Maria, and Josh White. "Va. Scientist Was Killed with Sword: Three Friends Interested in Occult and Witchcraft, Friends Say." *Washington Post*, December 14, 2001.

Good, Timothy. *Alien Liaison*. London, UK: Century, 1991.

Gorvett, Zaria. "You Are Surprisingly Likely to Have a Living Doppelganger." BBC.com, July 13, 2016. https://www.bbc.com/future/article/20160712-you-are-surprisingly-likely-to-have-a-living-doppelganger.

Gottlieb, Jeremy, and David Leech Anderson. "Robots: In the Beginning." The Mind Project. Accessed August 11, 2022. https://mind.ilstu.edu/curriculum/medical_robotics/robots_in_beginning.html.

Greenberg, Andy. "How the NSA Could Bug Your Powered-Off iPhone, and How to Stop Them." *Wired,* June 3, 2014. https://www.wired.com/2014/06/nsa-bug-iphone/.

Griffin, Andrew. "iPhone Has Secret Software That Can Be Remotely Activated to Spy on People, Says Snowden." *Independent,* January 21, 2015. http://www.independent.co.uk/life-style/gadgets-and-tech/news/iphone-has-secret-software-that-can-be-remotely-activated-to-spy-on-people-says-snowden-9991754.html.

"Group Claims Human Cloning Success." *Guardian,* December 27, 2002. https://www.theguardian.com/science/2002/dec/27/genetics.science.

Hennessy, Michelle. "Makers of Sophia the Robot Plan Mass Rollout Amid Pandemic." Reuters, January 24, 2021. https://www.reuters.com/article/us-hongkong-robot/makers-of-sophia-the-robot-plan-mass-rollout-amid-pandemic-idUSKBN29U03X.

Hollis, Heidi. *The Hat Man: The True Story of Evil Encounters.* Milwaukee, WI. Level Head Publishing, 2014.

Hornyak, Tim. "Robots at Home and Work: Where Are We Now?" Digital Arts, May 29, 2014. https://www.digitalartsonline.co.uk/features/hacking-maker/robots-where-are-we-now/.

Howard, Clark. "Your Smart TV Could be Spying on You." Clark.com, February 10, 2015. http://www.clark.com/your-smart-tv-spying-you.

Hsu, Jeremy. "Archimedes' Flaming Death Ray Was Probably Just a Cannon, Study Finds." *Christian Science Monitor,* June 29, 2010. https://www.csmonitor.com/Science/2010/0629/Archimedes-flaming-death-ray-was-probably-just-a-cannon-study-finds.

Kerner, Nigel. *Grey Aliens and the Harvesting of Souls: The Conspiracy to Genetically Tamper with Humanity*. Rochester, VT: Bear & Company, 2010.

Kerner, Nigel. *The Song of the Greys.* London, UK: Hodder & Stoughton, 1997.

Kleeman, Jenny. "The Race to Build the World's First Sex Robot." *Guardian,* April 27, 2017. https://www.theguardian.com/technology/2017/apr/27/race-to-build-world-first-sex-robot.

Kota, Sridhar, and Tom Kalil. "Developing the Next Generation of Robots." The White House—President Barack Obama, June 24, 2011. https://obamawhitehouse.archives.gov/blog/2011/06/24/developing-next-generation-robots.

Lamont, Tom. "I'll Do the First Human Head Transplant." *Guardian,* October 3, 2015. https://www.theguardian.com/science/2015/oct/03/will-first-human-head-transplant-happen-in-2017.

Marchetti, Victor. *CIA: The Cult of Intelligence.* New York: Dell, 1989.

Marks II, Robert J. "The Case for Killer Robots." Discovery.org, January 7, 2020. https://www.discovery.org/b/the-case-for-killer-robots/.

Maugham, Tim. "Meet Zoltan, the Presidential Candidate Who Drives a Coffin." BBC.com, November 29, 2015. https://www.bbc.com/future/article/20151127-meet-zoltan-the-strangest-candidate-running- for-president.

Mayor, Adrienne. "The World's First Robot." Wonders & Marvels, March 2012. https://www.wondersandmarvels.com/2012/03/the-worlds-first-robot-talos.html.

McDowell, Natasha. "Dutch Clone Claimed—But No Proof." *New Scientist,* January 6, 2003. https://www.newscientist.com/article/dn3230-dutch-clone-claimed-but-no-proof/.

Meghan, Neal. "Is a Human 'Barcode' on the Way?" *Daily News,* June 1, 2012. http://www.nydailynews.com/news/national/human-barcode-society-organized-invades-privacy-civil-liberties-article-1.1088129.

Mitchell, Alanna, Simon Cooper, and Carolyn Abraham. "Strange Cluster of Microbiologists' Deaths under the Microscope." *Globe and Mail,* May 4, 2002.

Mizokami, Kyle. "The U.S. Army Expects to Field Cyborg Soldiers by 2050." *Popular Mechanics,* November 26, 2019. https://www.popularmechanics.com/military/research/a29963287/us-army-cyborgs/.

Morris, Andrea. "Prediction: Sex Robots Are the Disruptive Technology We Didn't See Coming." September 25, 2018. *Forbes,* September 25, 2018. https://www.forbes.com/sites/andreamorris/2018/09/25/prediction-sex-robots-are-the-most-disruptive-technology-we-didnt-see-coming/?sh=73f07f5e6a56.

Moye, David. "This Real-Life Robocop Is on the Case at a DubaiShopping Mall." HuffPost.com, May 24, 2017. https://www.huffpost.com/entry/robocop-mall-cop_n_5925d8b4e4b062f96a336c3c.

Mullen, Leslie. "Cyborg Astronaut Space Race Heats Up." Space.com, April 8, 2010. https://www.space.com/8169-cyborg-astronaut-space-race-heats.html.

"NASA Takes a Cue from Silicon Valley to Hatch Artificial Intelligence Technologies." NASA.gov, November 19, 2019. https://www.nasa.gov/feature/goddard/2019/nasa-takes-a-cue-from-silicon-valley-to-hatch-artificial-intelligence-technologies.

Newling, Dan. "Britons 'Could Be Microchipped Like Dogs in a Decade.'" *Daily Mail,* October 30, 2006. http://www.dailymail.co.uk/news/article-413345/Britons-microchipped-like-dogs-decade.html.

Olla, Akin. "A Dystopian Robo-Dog Now Patrols New York City. That's the Last Thing We Need." *Guardian,* March 2, 2021. https://www.theguardian.com/commentisfree/2021/mar/02/nypd-police-robodog-patrols.

O'Meara, Sarah. "Will China Lead the World in AI by 2030?" *Nature,* August 21, 2019. https://www.nature.com/articles/d41586-019-02360-7.

Pain, Stephanie. "Blasts from the Past: The Soviet Ape-Man Scandal." *New Scientist,* August 20, 2008. https://www.newscientist.com/article/mg19926701-000-blasts-from-the-past-the-soviet-ape-man-scandal/.

Parker, Calvin. *Pascagoula—The Closest Encounter: My Story*. Privately printed, 2018.

Parviaininen, Janna, and Mark Coeckelbergh. "The Political Choreography of the Sophia Robot: Beyond Robot Rights and Citizenship to Political Performances for the Social Robotics Market." *AI & Society* 36, 715–724 (2021). https://link.springer.com/article/10.1007/s00146-020-01104-w. 2021.

Pellerin, Cheryl. "DARPA Provides Groundbreaking Bionic Arms to Walter Reed." U.S. Department of Defense, December 23, 2016. https://www.defense.gov/Explore/News/Article/Article/1037447/darpa-provides-groundbreaking-bionic-arms-to-walter-reed/.

Penniston, Jim, and Gary Osborn. *The Rendlesham Enigma: Book 1: Timeline.* Privately printed, 2019.

Petoe, Matthew, Ayton Lauren, and Shivdasani. "Artificial Vision: What People with Bionic Eyes See." The Conversation, August 16, 2017. https://theconversation.com/artificial-vision-what-people-with-bionic-eyes-see-79758.

Petrella, Stephanie, Chris Miller, and Benjamin Cooper. "Russia's Artificial Intelligence Strategy: The Role of State-owned Firms." *Orbis,* December 2020.

Randle, Kevin D., and Donald R. Schmitt. *UFO Crash at Roswell.* New York: Avon Books, 1991.

Randles, Jenny. *The Pennine UFO Mystery*. St. Albans, UK: Granada Publishing, Ltd., 1983.

———. *The Truth behind Men in Black*. New York: St. Martin's Paperbacks, 1997.

Rayne, Elizabeth. "Could Real-Life Cyborg Armies Be Marching Straight Out of Sci-Fi by 2050?" SyFy.com, December 4, 2019. https://www.syfy.com/syfywire/could-there-be-cyborg-armies-by-2050.

Redd, Nola Taylor. "What Is Wormhole Theory?" Space.com, October 21, 2017. https://www.space.com/20881-wormholes.html.

Redfern, Nick. *The Black Diary*. Bracey, VA: Lisa Hagan Books, 2018.

———. Interviews with Daniel M. Salter, March 13, 2003, and August 8, 2003.

———. Interview with Mac Tonnies, March 14, 2004.

———. Interview with Mac Tonnies, September 9, 2006.

———. Interview with Mac Tonnies, July 7, 2009.

———. *Men in Black*. Bracey, VA: Lisa Hagan Books, Bracey, 2015.

———. *The Real Men in Black*. Wayne, NJ: New Page Books, 2011.

———. *Secret History: Conspiracies from Ancient Aliens to the New World Order.* Canton, MI: Visible Ink Press, 2014.

Rees, Martin. "Why Man Turned His Back on the Moon: Exactly 50 Years after Humanity's Greatest Adventure, the Astronomer Royal's Compelling Insight into Why Cyborg Astronauts Will Make the Next Great Leap." *Daily Mail,* July 10, 2019. https://www.dailymail.co.uk/news/article-7266917/Astronomer-Royal-cyborg-astronauts-make-great-leap.html.

Rempfer, Kyle. "Cyborg Warriors Could be Here by 2050, DoD Study Group Says." *Army Times,* November 27, 2019. https://www.armytimes.com/news/your-army/2019/11/27/cyborg-warriors-could-be-here-by-2050-dod-study-group-says/.

"Report of the Select Committee on Assassinations of the U.S. House of Representatives." http://www.archives.gov/research/jfk/select-committee-report/. 2014.

Ruppert, Barb. "Robots to Rescue Wounded on Battlefield." U.S. Army, November 22, 2010. https://www.army.mil/article/48456/robots_to_rescue_wounded_on_battlefield.

Salge, Christoph. "Asimov's Laws Won't Stop Robots from Harming Humans, So We've Developed a Better Solution." *Scientific American,* July 11, 2017. https://www.scientificamerican.com/article/asimovs-laws-wont-stop-robots-from-harming-humans-so-weve-developed-a-better-solution/.

Salter, Daniel M. *Life with a Cosmos Clearance.* ? Flagstaff, AZ: Light Technology Publishing. June 1, 2003.

Sato, Hirotaka, et al. "Remote Radio Control of Insect Flight," *Frontiers in Integrative Neuroscience,* October 5, 2009. https://doi.org/10.3389/neuro.07.024.2009.

Saul, Heather. "Reversing Walking Corpse Syndrome: Cotard's Syndrome Trigger Found—And It's a Household Cold Sore Cream." *Independent,* October 18, 2013. https://www.independent.co.uk/news/science/reversing-walking-corpse-syndrome-cotard-s-syndrome-trigger-found-and-it-s-household-cold-sore-cream-8888670.html.

Scharr, Jillian. "Wormhole Is Best Bet for Time Machine, Astrophysicist Says." Live Science, August 25, 2013. https://www.livescience.com/39159-time-travel-with-wormhole.html.

Seaburn, Paul. "Physicist Describes Wormhole That Will Make Time Travel Possible." Mysterious Universe, November 19, 2017. https://mysteriousuniverse.org/2017/11/physicist-describes-wormhole-that-will-make-time- travel-possible/.

Seaburn, Paul. "Woman Injects 3.5-Million-Year-Old Bacteria to Stay Young." Mysterious Universe. October 25, 2017. https://mysteriousuniverse.org/2017/10/woman-injects-3-5-million-year-old-bacteria-to-stay-young/.

Shashkevich, Alex. "Stanford Researcher Examines Earliest Concepts of Artificial Intelligence, Robots in Ancient Myths." Stanford News, February 28, 2019. https://news.stanford.edu/2019/02/28/ancient-myths-reveal-early-fantasies-artificial-life/.

Snyder, Michael. "After the Government Microchips Our Soldiers, How Long Will It Be before They Want to Put a Microchip in YOU." End of the American Dream, May 8, 2012. http://endoftheamericandream.com/archives/after-the-government-microchips-our-soldiers-how-long-will-it-be-before-they-want-to-put-a-microchip-in-you.

South, Todd. "War with Robots: How Battle Bots Will Define the Future of Ground Combat." *New York Times.* February 13, 2020. https://www.armytimes.com/news/your-army/2020/02/13/war-with-robots-how-battle-bots-will-define-the-future-of-ground-combat/.

Steiger, Brad. *Conspiracies and Secret Societies.* Detroit: Visible Ink Press, 2013.

———. "Three Tricksters in Black." *Saga's UFO Report,* Winter 1974.

Strieber, Whitley. *Communion: A True Story*. New York: Beach Tree Books, 1987.

Sukheja, Bhevya. "Elon Musk Reveals Ambitious Plans to Get to Mars Seven Years Ahead of NASA." RepublicWorld.com, February 2, 2021. https://www.republicworld.com/technology-news/science/elon-musk-reveals-ambitious-plans-to-get-humans-to-mars-seven-years-ahead-of-nasa.html.

Swancer, Brent. "The Creepy World of Sophia the Robot." Mysterious Universe, February 11, 2021. https://mysteriousuniverse.org/2021/02/the-creepy-world-of-sophia-the-robot/.

Teitel, Amy Shira. "Future Spacemen May Not Be Men at All: How Robots, Cyborgs and Mutants Will (and Won't) Be Better Astronauts than Us." VICE, July 11, 2011. https://www.vice.com/en/article/xyy787/will-future-spacemen-be-men-at-all.

Tingley, Brett. "BDSM, Sex Robots, and the Three Laws of Robotics." Mysterious Universe. Accessed August 11, 2022. https://mysteriousuniverse.org/2018/10/bdsm-sex-robots-and-the-three-laws-of-robotics/.

———. "Robot Gains National Citizenship and Demands Equal Rights with Humans." Mysterious Universe. Accessed August 11, 2022. https://mysteriousuniverse.org/2017/10/robot-gains-national-citizenship-and-demands-equal-rights-with-humans/.

Tonnies, Mac. *The Cryptoterrestials*. San Antonio, TX: Anomalist Books, 2010.

United States Air Force. *The Roswell Report: Case Closed.* U.S. Government Printing Office, 1997.

United States Air Force. *The Roswell Report: Fact versus Fiction in the New Mexico Desert.* U.S. Government Printing Office, 1994.

U.S. Department of State. "Strategic Defense Initiative (SDI) 1983." Accessed August 11, 2022. https://2001-2009.state.gov/r/pa/ho/time/rd/104253.htm.

"US Family Gets Health Implants." BBC News. BBC, May 11, 2002. http://news.bbc.co.uk/2/hi/health/1981026.stm.

Weatherly, David. *Black Eyed Children*. NV: Leprechaun Press, 2017.

Weller, Chris. "The First 'Robot Citizen' in the World Once Said She Wants to 'Destroy Humans.'" Inc.com, October 26, 2017. https://www.inc.com/business-insider/sophia-humanoid-first-robot-citizen-of-the-world-saudi-arabia-2017.html.

Wilkins, Alasdair. "Cyborg Astronauts Will Be Needed to Colonize Outer Space." Gizmodo, December 16, 2015. https://gizmodo.com/cyborg-astronauts-will-be-needed-to-colonize-outer-spac-5640444.

Wilmut, Ian. "John Clark: Pioneering Scientist Whose Entrepreneurial Skills Paved the Way for Dolly the Sheep." *Guardian*, August 25, 2004.

Wood, Robert M., and Nick Redfern. *Alien Viruses*. Rochester, NY: Richard Dolan Press, 2013.

Wood, Ryan. *Majic Eyes Only*. Broomfield, CO: Wiid Enterprises, 2005.

Zetter, Kim. "April 13, 1953: CIA OKs MK-Ultra Mind-Control Tests." Wired, April 13, 2010. http://www.wired.com/2010/04/0413mk-ultra-authorized/.

Index

Note: (ill.) indicates photos and illustrations.

B

C

D

N

R

S

T